Communication in a Virtual Organization

Managerial Communication Series

By **Sandra D. Collins**
University of Notre Dame

Series Editor: **James S. O'Rourke, IV**
University of Notre Dame

THOMSON
™
SOUTH-WESTERN

Australia · Canada · Mexico · Singapore · Spain · United Kingdom · United States

To my family:
Pam, Colleen, Molly, and Kathleen.
And to my colleagues:
Carolyn, Sandra, Cynthia, and Renee.
Thanks.
JSO'R, IV

To Jarrod and Garett.
And with thanks to Jim and Ron.
SDC

THOMSON

SOUTH-WESTERN

Communication in a Virtual Organization, Managerial Communication Series

James S. O'Rourke, IV, series editor; Sandra D. Collins, author

Editor-in-Chief:
Jack Calhoun

VP/Team Leader:
Melissa Acuña

Acquisitions Editor:
Jennifer Codner

Developmental Editor:
Taney Wilkins

Marketing Manager:
Larry Qualls

Production Editor:
Robert Dreas

Manufacturing Coordinator:
Diane Lohman

Compositor:
Lachina Publishing Services

Printer:
Transcontinental Printing, Inc.
Louiseville, Quebec

Sr. Design Project Manager:
Michelle Kunkler

Cover and Internal Designer:
Robb and Associates

Cover Illustration:
© Artville/Richard Cook

Library of Congress
Control Number:
2002111622

ISBN: 0-324-15256-6

AUTHOR BIOGRAPHIES

James Scofield O'Rourke, IV, is director of the Eugene D. Fanning Center for Business Communication at the University of Notre Dame, where he teaches writing and speaking. In a thirty-five year career, he has earned an international reputation in business and corporate communication. *Business Week* magazine again named him one of the "outstanding faculty" in Notre Dame's Mendoza College of Business. Professor O'Rourke has held faculty appointments in such schools as the United States Air Force Academy, the Defense Information School, the United States Air War College, and the Communication Institute of Ireland. He is a regular consultant to Fortune 500 and mid-size businesses, and is widely published in both professional journals and the popular press.

Sandra Dean Collins currently teaches management communication for the Mendoza College of Business at the University of Notre Dame. Her courses include business writing, speaking, listening and responding, and managing differences. She has also taught statistics and research methods for the university. She conducts team training for the Mendoza College of Business and local organizations and consults with small and mid-sized organizations on communication and team related issues. Her background includes a Ph.D. in Social Psychology and experience in sales, purchasing, and banking.

TABLE OF CONTENTS

FOREWORD

In recent years, for a variety of reasons, communication has grown increasingly complex. The issues that seemed so straightforward, so simple not long ago are now somehow different, more complicated. Has the process changed? Have the elements of communication or the barriers to success been altered? What's different now? Why has this all gotten more difficult?

Several issues are at work here, not the least of which is pacing. Information, images, events, and human activity all move at a much faster pace than they did just a decade ago. The most popular, hip new business magazine is named *Fast Company*. Readers are reminded that it's not just a matter of tempo, but a new way of living we're experiencing.

Technology has changed things, as well. We're now able to communicate with almost anyone, almost anywhere, 24/7 with very little effort and very little professional assistance. It's all possible because of cellular telephone technology, digital imaging, the Internet, fiber optics, global positioning satellites, teleconferencing codecs, high-speed data processing, online data storage, and . . . well, the list goes on and on. What's new this morning will be old hat by lunch.

Culture has intervened in our lives in some important ways. Very few parts of the world are inaccessible anymore. Other people's beliefs, practices, perspectives, and possessions are as familiar to us as our own. And for many of us, we're only now coming to grips with the idea that our own beliefs aren't shared by everyone and that culture is hardly value-neutral.

For a thousand reasons, we've become more emotionally accessible and vulnerable than ever before. You may blame Oprah or Jerry Springer for public outpouring of emotions, but they're not really the cause—they're simply another venue for joy, rage, or grief. The spectacle of thousands of people in London mourning at the death of Diana, Princess of Wales, took many of us in the United States by surprise. By the time the World Trade Center towers came down in a terrorist attack, few of us had tears left to give. Who could not be moved by images of those firefighters, laboring in the night, hoping against hope to find a soul still alive in the rubble?

The nature of the world in which we live—one that's wired, connected, mobile, fast-paced, iconically visual, and far less driven by logic—has changed in some not-so-subtle ways in recent days. The organizations that employ us and the businesses that depend on our skills now recognize that communication is at the center of what it means to be successful. And at the heart of what it means to be human.

To operate profitably means that business must now conduct itself in responsible ways, keenly attuned to the needs and interests of its stakeholders. And, more than ever, the communication skills and capabilities we bring to the workplace are essential to our success, at both the individual and the societal level.

So, what does that mean to you as a prospective manager or executive-in-training? For one thing, it means that communication will involve more than simple writing, speaking, and listening skills. It will involve new contexts, new applications, and new technologies. Much of what will affect the balance of your lives has yet to be invented. But when it is, you'll have to learn to live with it and make it work on your behalf.

The book you've just opened is the third in a series of six that will help you to do all of those things and more. Professor Sandra Collins, a social psychologist by training, explores *Commu-*

nication in a Virtual Organization. The conceptual framework she brings to the discussion will help you to understand how the compression of time and distance has altered work habits and collaboration. With the help of corporate communication executives and consultants, she offers current examples of global companies and local groups that illustrate the ways in which our work and lives have permanently changed.

In this series's first module, Professor Bonnie Yarbrough examines issues related to *Leading Groups and Teams.* She reviews the latest research on small group and team interaction and offers practical advice on project management, intra-team conflict, and improving results. In module two, Professor Carolyn Boulger offers a radical new view of technology's impact on our society and the organizations that employ us in *e-Technology and the Fourth Economy.* Everything from e-learning to virtual teams to our expectations of privacy in the workplace are offered up in a compact, readable format.

For the iconically challenged (I am one who thinks in words and phrases, not pictures), Professor Robert Sedlack (graphic arts) and Professor Cynthia Maciejczyk (communications) will explore *Graphics and Visual Communication for Managers.* If you've ever wondered how to transform words and numbers into pictures, this module can help. And for all of us who've ever tried to explain complex issues without success, either aloud or on paper, the message is simple: If you can't say it in a clear, compelling way, perhaps you can show them.

Professor Collins will also examine *Managing Conflict and Workplace Relationships* in a volume that may have more of a lasting impact on how (and with whom) we can work than any of the other titles. Her approach involves far more than dispute resolution or figuring out how limited resources can be distributed equitably among people who think they all deserve more. She shows us how to manage our own emotions, as well as those of others. Creative conflict, harmony, and synchronicity in the workplace are issues too many of us have avoided because we simply didn't understand them or didn't know what to say.

Finally, Professor Yarbrough will examine *International and Intercultural Communication,* looking both broadly and specifically at issues and opportunities that will seem increasingly important as the business world shrinks and grows more interdependent. As time zones blur and fewer restrictions are imposed on the global movement of capital, raw materials, finished goods, and human labor, people will cling fiercely to the ways in which they were enculturated as youngsters. Culture will become a defining characteristic, not only of peoples and nations, but of organizations and industries.

This is an interesting, exciting, and highly practical series of books. They're small, of course, intended not as comprehensive texts, but as supplemental readings, or as stand-alone volumes for modular courses or seminars. They're engaging because they've been written by people who are smart, passionate about what they do, and more than happy to share what they know. And I've been happy to edit the series, first, because these writers are all friends and colleagues whom I know and have come to trust. Secondly, I've enjoyed the task because this is really interesting stuff. Read on. There is a lot to learn here, new horizons to explore, and new ways to think about human communication.

James S. O'Rourke, IV
The Eugene D. Fanning Center
Mendoza College of Business
University of Notre Dame
Notre Dame, Indiana

Managerial Communication Series

Editor: James S. O'Rourke, IV

The **Managerial Communication Series** is a series of modules designed to teach students how to communicate and manage in today's competitive environment. Purchase only this module as a supplemental product for your Business Communication, Management, or other Business course, or purchase all six modules, packaged together at a discounted price for full coverage of Managerial Communication.

Leading Groups and Teams
Bonnie T. Yarbrough

1 Managerial Communication Series | Editor: James S. O'Rourke, IV

ISBN: **0-324-15254-X**

This text, written by Bonnie T. Yarbrough, reviews the latest research on small group and team interaction, and offers practical advice on project management, intra-team conflict, and improving results. It contains group and team worksheets, progress reports, and sample reporting instruments, as well as classroom discussion questions and case studies.

e-Technology and the Fourth Economy
Carolyn A. Boulger

2 Managerial Communication Series | Editor: James S. O'Rourke, IV

ISBN: **0-324-15255-8**

This text, written by Carolyn A. Boulger, offers a radical new view of technology's impact on what the author calls "The Fourth Economy," an economic model based entirely on minds in interaction. Technology's role in helping participants in the radically transformed landscape of the twenty-first century is not limited to the transmission and storage of text and data, but extends to the very ways in which people think about and create value.

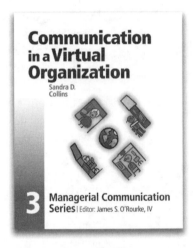

Communication in a Virtual Organization
Sandra D. Collins

3 Managerial Communication Series | Editor: James S. O'Rourke, IV

ISBN: **0-324-15256-6**

This text, written by Sandra D. Collins, explores the risks and opportunities open to those who work in new alliances, partnerships, and non-traditional business models. A look at both theory and practical application offers students and managers the chance to observe successful organizations in action.

Contact your local South-Western/
Thomson Learning Representative at
800-423-0563. Or visit the series Web site
at **http://orourke.swcollege.com** *for more*
product information and availability.

Graphics and Visual Communication for Managers
Robert P. Sedlack, Jr.
Cynthia S. Maciejczyk

4 Managerial Communication Series | Editor: James S. O'Rourke, IV

ISBN: **0-324-16178-6**

This text, written by Robert P. Sedlack, Jr., and Cynthia S. Maciejczyk, offers some practical and useful advice on how to work with graphics and visuals in reports, briefings, and proposals. It also offers direct instruction on how to integrate graphic aids into spoken presentations and public speeches. If you can't say it or write it clearly, you may be able to show it. Dozens of illustrations, drawings, and graphs are included.

Managing Conflict and Workplace Relationships
Sandra D. Collins

5 Managerial Communication Series | Editor: James S. O'Rourke, IV

ISBN: **0-324-15257-4**

Learn what social scientists and business executives now know about conflict, personality style, organizational structure, and human interaction. This text, written by Sandra D. Collins, examines the most successful strategies for keeping your edge and keeping your friends. Practical forms, instruments, and applications are included.

International and Intercultural Communication
Bonnie T. Yarbrough

6 Managerial Communication Series | Editor: James S. O'Rourke, IV

ISBN: **0-324-15258-2**

This text, written by Bonnie T. Yarbrough, examines the basis for culture, reviewing the work of social scientists, cultural anthropologists, and global managers on this emerging topic. Definitions of culture, issues of cultural change, and how cultures adapt are included, along with practical examples, case studies, and illustrations of how cultural issues are managed both domestically and internationally.

INTRODUCTION

When people are asked if they have any personal experience working in a virtual organization, they often respond by inquiring, "What is a virtual organization?" It's a good question because the definition changes depending on whom you talk to. In general, a virtual organization isn't a distinct business model or way of organizing a business, rather it's a way of working together—without being together—that can take many forms. In a virtual organization, some or all of the business relationships are maintained entirely or in part through electronically mediated communication.

With this very broad definition in mind, most people could easily find ways to describe their traditional working environments as virtual organizations. After all, for decades people have had extensive dealings, via the phone or regular mail, with co-workers in another building or across town or even out of state, or with distant clients that they rarely, if ever, interacted with face to face.

The difference today is the amount and complexity of work that can be done from distant locations without necessarily requiring a real-time interaction (although even real-time interaction can be handled electronically with videoconferencing). Previously, the need to access file cabinets full of information, use forms, or transfer information to others inside or outside the organization prohibited much work from being done off-site. But today's technology enables people to access databases, transfer documents, complete online forms, and juggle much more from any location and without interacting with a co-worker located in the office. Many jobs can be completed entirely from a remote location by means of electronic communication.

Organizations benefit in numerous ways from implementing virtual work environments. Companies save on reduced real estate costs and lowered absenteeism and turnover. Organizations are able to affordably tap into the knowledge of experts no matter where they are located. Candidates for employment are attracted to the flexibility of working off-site. And research on telecommuting indicates that worker productivity increases when people are permitted to work from home.

Despite these advantages, at the present time many organizations do not offer any formal virtual work arrangements. And even in those organizations that do, most employees that work from remote locations, including those that work from their homes, still go into the office part of the week. Often this time spent in the traditional office is the result of a choice made by the manager or the employee, rather than a requirement for the completion of job-related tasks.

The main hindrance to the growth of off-site work arrangements is hesitancy on the part of management. This is not entirely surprising because working virtually requires a management style that flies in the face of traditional methods. Managers are uneasy about how to replace their line-of-sight methods of management with other techniques that they are equally confident will work. Managers often hold a Theory X perspective of employees, particularly clerical staff, and believe that employees won't work unless someone is watching to make sure they do. Traditional managers use methods of evaluation that focus on attendance, punctuality, and attitude, which are difficult to assess in an off-site worker.

Virtual work environments represent change for managers and employees. And as with any change in general, preparing for it can make the transition smoother. Organizations that spend time training their managers to manage off-site workers, use rigorous screening processes to select off-site workers, and offer remote workers training and support have successful programs and realize true benefits. The goal of this brief book is to provide managers, remote workers, and virtual team members with an overview of the issues and solutions associated with virtual work environments, in whatever form they take.

Chapter 1, A Coming of Age, describes the forms that virtual work environments may take. It discusses the advantages and disadvantages of the virtual workplace for the employer and employee and gives the descriptive statistics on the prevalence and growth of working from home and the demographics of the typical telecommuter.

Chapter 2, The Communication Process, explains the process of communication and describes communication barriers frequently encountered in organizations. The effects of communicating electronically rather than face to face are discussed along with suggestions for overcoming the communication barriers in virtual work environments.

Chapter 3, Technologies for Virtual Organizations, reviews the technologies available for communicating in a virtual organization and offers suggestions for using them effectively. New communication technologies hit the market regularly, so this chapter offers a review of general categories of technologies, rather than particular products.

Chapter 4, The Medium and the Message, explores theories of media choice, including media richness theory and social presence theory. How the medium used to deliver a message affects the interpretation of that message is discussed.

Chapter 5, Managing Virtual Work Environments, discusses the best practices associated with managing off-site workers and working as part of a virtual team. Implementing virtual work environments successfully requires preparation from managers and employees. This chapter offers suggestions for selecting remote workers, training and evaluating them, and keeping them connected.

The trend toward implementing virtual work arrangements will certainly continue to expand as technological advances make it even easier and as more information becomes available on how to do so successfully. Organizations that take advantage of virtual work environments for saving costs, increasing productivity, and attracting talent may force even resistant organizations to seek the competitive advantage gained by a distributed workforce. The key to successful implementation is being prepared and knowing what to expect. This book is a good place to start.

1 A COMING OF AGE

"I can't figure out how people could operate a business today without virtual teams," says Elizabeth Allen, Vice President of Corporate Communication at Dell Computer. "We're a global company." Case in point: It was 10:05 a.m. at Dell's corporate headquarters in Round Rock, Texas, when Allen made these comments. At 10:30 she had a teleconference scheduled with members of a global virtual team about the annual report. Dell has employees all over the world, and they communicate primarily by e-mail, sometimes by teleconference, and rarely by videoconference. Allen tries to bring members of her team to the United States for a face-to-face visit occasionally as well.[1]

Many people share Allen's thoughts about virtual teams. Advanced technology has opened up the globe to organizations and made the barriers of time and distance nearly irrelevant. Organizations have easy electronic access to people and their expertise, anytime, anywhere. In addition to making people more accessible, technology has opened the gates to a wealth of information. With a few keystrokes, anyone can access a tremendous amount of information on almost any topic. To survive in today's competitive marketplace, organizations must know how to manage information and knowledge effectively.

As we move into the Digital Age, those organizations that are skilled both at using technology to create communication networks and at harvesting and sharing knowledge will have a competitive advantage. Using communication networks and information effectively means an increased ability to respond rapidly to unpredictable and sometimes chaotic markets.

To meet demands for responsiveness and adaptability, many organizations are creating virtual work environments. Virtual workplaces increase an organization's ability to adapt in several ways. Global virtual teams allow organizations to span time zones using a tag-team approach to getting things done that essentially follows the sun around the globe. The network support team for Accenture, formerly

Virtual Wisdom ▼

You have no choice but to operate in a world shaped by globalization and the information revolution. There are two options: adapt or die.

—Andrew Grove, Intel Corporation[2] ▲

Anderson Consulting, works on networking problems during the day, then passes them to a team in the next designated time zone, allowing work to continue around the clock.[4] Virtual teams give organizations access to the best people to deal with a problem, no matter where they live. The National Aeronautics and Space Administration (NASA) uses virtual teams made up of members from diverse locations and organizations that float in and out of the team as their expertise is needed.[5] Organizations can respond more readily to the needs of clients by placing off-site employees near or with them. For example, in 1995 Xerox closed a Massachusetts location and sent thousands of its employees to work at clients' facilities or to work from their homes located near clients.[6]

WHAT IS A VIRTUAL WORKPLACE?

There is currently no narrow, agreed-upon definition of a virtual workplace. A virtual organization is not a single organizational design. Instead, the term refers to a number of ways of working together, while apart. A virtual organization can be broadly defined as an organization that forms, and/or maintains, some or all of its internal or external relationships electronically.[7] Its work products are electronic rather than paper.[8] This broad definition of *virtual organization* includes organizations whose relationships with clients, customers, vendors, consultants, or joint venture partners are maintained virtually.[9] For example, Levi Strauss & Company uses a quick response system that links its distribution centers and retail outlets. In this way, Levi knows how many pairs of jeans were sold in each of its authorized retailers daily.

Researchers have proposed numerous typologies to describe the forms virtual organizations take. *Permanent virtual organizations* are designed to have some or part of the ongoing business activity performed virtually as a means of generating revenue and/or saving on cost. *Virtual teams* are groups of people that work together for a common goal across boundaries of time and distance. Virtual teams can be intra-organizational or inter-organizational, with their members selected on the basis of their expertise from all over the globe. Virtual teams often have a temporary nature and are disbanded as their goals are met. With *virtual operations,* certain duties are completed off-site. For example, service agents who dial into the Internet can answer all incoming customer service calls anywhere. *Temporary virtual organizations* are formed to take advantage of a specific market opportunity and are disbanded when the opportunity has passed.[11]

Individuals that work in virtual environments are often referred to as telecommuters or teleworkers because they work away from an employer's place of business and rely on communication technologies, such as the Internet or telephone, to communicate. In many cases, telecommuters work from home, but not always. Organizations with offices located in large metropolitan areas, such as Charles Schwab and Sun Microsystems, offer their telecommuters the opportunity to work from telework centers.[13] These are offices located in areas more accessible to teleworkers than the downtown office. They can take the form of satellite offices, where all employees are from the same organization, or work-centers that are shared by people from a number of organizations. The available offices are not permanently assigned. With this practice, sometimes called *hoteling,* employees can reserve an office for the day and take their belongings home at night. The offices usually have some administrative support staff to assist workers.

> **Virtual Wisdom** ▼
>
> *When we talk "virtual organizations," we are really talking about dramatically different ways of organizing capital, technology, information, people, and other assets than we utilized in the past.*
>
> —Dana Mead, retired chairman and CEO
> Tenneco, Inc.[12] ▲

Gil Chiquito, Audit CPA for Arthur Andersen in Dallas, Texas, had mixed reactions to his hoteling experience. "The top benefit of hoteling is that it allows you to get away from it all by finding quiet corners or floors to set up your workspace." Chiquito enjoyed choosing spots away from high traffic areas and his supervisor's line of vision to reduce interruptions. However, he notes the difficulty of building relationships when hoteling. "Because hoteling allows individuals to sit virtually anywhere in the office, you are constantly around new faces."[14]

Telecommuters are, to some extent, location independent and thus not limited to working from home or work centers. In the case of what is called extreme telecommuting, people work from far-flung regions. Consider Paolo Conconi, owner of MPS Electronics in Hong Kong and extreme telecommuter. While living in Hong Kong, Conconi fell in love with a woman from Jakarta. He began visiting her for long periods of time and found that he was able to continue conducting business during his visits. It occurred to him that if he could do that over the length of a long visit, he could do it permanently. He moved from Hong Kong to Jakarta, and later the couple moved to Bali, Indonesia. Conconi now conducts his work throughout China and Europe from poolside at his home in Bali. Conconi likes the pace of life there and enjoys having a villa with a pool for half of the rent he paid for his apartment in Hong Kong.[15]

THE GROWTH OF THE VIRTUAL WORKPLACE

These various forms of virtual organizations and telecommuting are growing in popularity. An estimated 23.6 million Americans telecommuted in 2001. This is an increase from just 4 million in 1990, and the number is projected to reach 38 million by 2006.[16]

The typical teleworker is a college-educated male, about 34 to 55 years old. Almost a third of teleworkers are Generation Xers, between the ages of 18 and 34, while about half are between the ages of 35 and 54. The majority of them (76%) are white and most (74%) earn more than $40,000 per year. Many (44%) earn more than $75,000 per year. Despite some concerns about job security, teleworkers are likely to be satisfied with their jobs. Most of them do not telecommute every day, as you can see in Table 1-1.[17] On average, in 1999 telecommuters worked nine days per month from a location other than the central office.

The kind of work that a teleworker does ranges from professional to managerial to clerical to service to technical. Nearly half of all telecommuters work for companies with 250 or more employees. Over half of telecommuters (57%) work for private, for-profit companies, 19% are self-employed, 13% work for nonprofit organizations, and 11% work for the government.[18]

In the late 1990s, approximately 25% of companies formally offered telecommuting as an option for their employees. This number may be a low estimate, however, because managers tend to report higher levels of telework opportunities when informal telecommuting is included. Larger companies are much more likely to have formal teleworking arrangements. Currently, 70% of companies with more than 5,000 employees have telecommuters.[19]

A survey conducted in the Houston, Texas, area by Joanne H. Pratt Associates, a firm that does research on telecommuting, indicates that the opportunities for telework vary by industry, with larger percentages of the mining, wholesale, and services industries offering telework options than retail, construction, and manufacturing (see Table 1-2).

Many employees would like to telecommute all or part of the time but are not offered the option by their employers. Of course, not all jobs are appropriate for telework; for example, production and quality workers in a manufacturing environment, researchers that use expensive lab equipment, or cashiers in a retail operation must perform their duties on-site. The U.S. government has been slower to adopt the practice of telecommuting than the private sector because many agencies deal with classified or confidential documents that cannot be taken home. Still, according to Representative Frank Wolf, a Virginia Republican and supporter of telecommuting, of the 1.8 million employees of the executive agencies, 45% to 60% could do their jobs off-site.[20]

Table 1-1   Number of Days Workers Telecommute per Week

Number of Days Teleworking per Week	Percentage of Teleworkers
1	30%
2	17%
3	12%
4	5%
5	17%
6	7%
7	10%

Source: Carl E. Van Horn and Duke Storen, "Telework: Coming of Age? Evaluating the Potential Benefits of Telework," *Telework and the New Workplace of the 21st Century*, 2000. Available: *http://www.dol.gov/asp/telwork/p1_1.html.*

Table 1-2   Percentage of Industry That Currently Permits or Is Considering Telework

Industry	Size of Sample	Percentage of Industry That Currently Permits Telework	Percentage of Industry That Is Considering Offering Telework
Wholesale	10	51%	44%
Mining (oil & gas)	22	46%	22%
Services	56	37%	37%
Transportation	10	26%	26%
Finance, Insurance, Real Estate	13	25%	8%
Construction	33	17%	28%
Manufacturing	10	14%	29%
Retail	7	0%	0%

Source: Joanne H. Pratt, "Why Aren't More People Teleworking?" *Transportation Record,* 1607, p. 168. Available: *http://www.joannepratt.com/images/whyarentmorepeople.pdf.*

DRIVING FACTORS IN THE GROWTH OF VIRTUAL WORK ENVIRONMENTS

Why are people that have the option choosing to telecommute? Why are so many employers making the option available and so many others considering it? What drives this growth trend? In the next several pages, we'll explore the motivations for employees and employers to participate in virtual work arrangements.

ADVANCED TECHNOLOGY

Many of the factors driving the trend toward telecommuting are both societal and economic. The process of moving from a technologically simple society to a more complex one is called sociocultural evolution. As societies become more technologically advanced and complex, the culture and lifestyles of their members are dramatically affected.

The humble plow is a great example of how a technological advancement can radically change a culture. Until about 6,000 years ago, it was necessary for almost everybody in a society to participate in food production because it was so labor-intensive. Then the plow was invented. Now a single farmer could produce much more food than his or her family could consume. As a result of increased food production capabilities, populations grew and many members of society were free to do things other than farm.[21]

As we move into the Digital Age, we can observe how advanced technology has changed, and continues to change, the nature and environment of work. We feel the impact of these changes in every aspect of our jobs, from how we seek employment and market ourselves to potential employers to how long we stay with an employer before moving on. We post resumes online rather than print and mail a hundred copies. We e-mail people in the office next to ours instead of leaning in the doorway to chat. We take our laptops on vacation so that we can keep in touch with the office, and, when we feel restless, we search the Internet for interesting new career opportunities.

Advances in technology are the most important contributors to the growth of telecommuting. The driving force behind the current attention focused on virtual work environments is the development and availability of affordable communication technologies. Of course, people have informally telecommuted for as long as telephones have been widely available. They have taken work home with them for years and called co-workers or clients. But more recent high-tech appliances, such as the personal computer, make it possible for people to accomplish much more complex tasks than they previously were able to do from home. Not so long ago computers were very large and extremely expensive, and the thought of owning a computer seemed impractical and not particularly desirable. Ken Olson, president of Digital Equipment Company in 1977, once said, "There is no reason anyone would want a computer in their home."[22] Today, home computers are so powerful and affordable that you can purchase a computer that will do everything your computer at work can do for about $700. A survey conducted by National Science Foundation in 1999 revealed that 54 percent of Americans owned home computers. [23]

TYPES OF WORK AVAILABLE

As we become primarily a post-industrial society, we see the number of manufacturing jobs shrink and the number of service jobs increase. After the Industrial Revolution, as technology improved and processes became further industrialized, fewer blue-collar workers were needed to produce the same amount of product. Industrialization created high levels of productivity, which led to greater prosperity. This, in turn, led to an increase in the demand for all sorts of services and the creation of more white-collar jobs. Today, even in manufacturing companies, white-collar workers often outnumber blue-collar laborers.[24]

Post-industrial societies subsist primarily from the production of services and knowledge, especially technical knowledge. Knowledge and information are products that may be purchased, sold, or leased like any other product.

In a knowledge economy, the means of production shifts from machines to "gray matter" and an accompanying personal computer. Fortunately for those interested in working off-site, both of these travel fairly well. In past societies, workers were tied to the seasons, the herd, the soil, or the machine, but now work can be performed nearly anywhere. After the attacks of September 11, 2001, the employees of Lightreading.com, an online news service company with offices only a half mile from the World Trade Center, began working at home. According to Stephen Saunders, the company founder, it was easy to convert Lightreading.com to a virtual organization. "Everything we do is online anyway," he said. For the first time, there is an alternative: The work can come to the worker, rather than the worker going to the work.[25]

CHANGES IN ORGANIZATIONAL STRUCTURES

After the Industrial Revolution, larger cities and larger companies resulted in the emergence of large, impersonal bureaucracies. Bureaucratic organizations, of course, were perceived as efficient ideals: necessary but not well liked. Efficiency was increased by the attempted removal of the personal aspect of work. Organizations were to be as mechanical as their machines.[26]

Today it is common for teams of individuals—sometimes not even in the same location—to take a process from beginning to end, rather than to perform individual work that focuses on one element of the task.

—Michael Verespej, business writer[27] ▲

Contributing to this impersonal feeling, industrialization also led to specialization, changing the relationship of those at work to the work itself. Workers could produce more if they were required to complete just one small part of the production process. They had no connection to the process as a whole. Unlike true artisans, the modern blue-collar worker is separated from the end product of his or her labor.

Bureaucracies are well-known for the red tape one typically encounters in dealing with them. In very large, complex organizations, a certain amount of red tape may be unavoidable. But today's workers are more highly educated than ever before and are accustomed to having job-related information quickly available. Even without higher education, Americans have the ability to be well-informed. The advanced technology of post-industrialism allows ordinary people instant access to extraordinary amounts of information. At the time of this writing, an Internet search of the term *virtual organizations* produced 39,900 hits. People today have little patience for slow-moving bureaucracies that appear to be constrained by meaningless rules, outmoded policies, and inefficient procedures.

The specialists and technical experts in demand today are used to having access to information in a way that is not common in traditional, hierarchical organizations. In general, these workers prefer to be more autonomous and to work in what are sometimes called collectivist organizations that recognize the value of collaboration and networking rather than relying on pure authority, as in a bureaucratic organization (see Table 1-3). Collectivist organizations have flatter organizational charts and are more egalitarian in nature than the hierarchies of bureaucracy. Because the emphasis is on egalitarianism and collaboration, rather than control, with organizations of this sort, it is possible to complete work outside the physical presence of superiors in the organization.

URBANIZATION

After the inventions of the Industrial Revolution mechanized farming in the United States, only a small number of farmers were needed to produce food. People who were no longer needed for farming moved to the cities to find work in one of many industries, thereby creating densely populated urban areas.

At the famous Corporate CEO Citizenship Breakfast at the White House in 1995, President Clinton quoted one of the mill's hourly workers, who observed that in the old days his bosses only hired him "from the neck down," but now he works "from the neck up."

—Dana Mead, buisness writer[28] ▲

Table 1-3   Characteristics of Bureaucratic and Collectivist Organizations

Bureaucratic Organizations	Collectivist Organizations
Maximum division of labor	Minimum division of labor
Maximum specialization of jobs	Diffusion of expertise
Emphasis on hierarchies	Egalitarianism
Authority in officeholders	Subordinate participation
Formalized, fixed rules	Primacy of ad hoc decisions
Worker motivation through direct supervision	Worker motivation through personal appeals
Impersonality as the ideal of social relations in organization	Comradeship as ideal of social relations in organizations

Source: Alex Thio, *Sociology,* 5th ed. New York: Addison Wesley Longman, 1998.

Housing shortages quickly developed in large cities, and other inconveniences made urban life unappealing. So, the more affluent people in society moved to the suburbs. This was facilitated by a government subsidy for highways, making it possible to commute. The number of wage earners fleeing the central cities of the United States increased rapidly during the twentieth century. In 1940, only 20% of Americans lived in the suburbs. By 1990, the figure increased to nearly 45%. Among the consequences of this rush to the suburbs has been an increase in commuter traffic. Brent Daniel, who works for Sun Microsystems, avoids a 75-minute commute once a week to the company's Mountain View office by working from the company's call center located 15 minutes from his house.[29] Martin Gertel, audit liaison for the U.S. Department of Transportation, often avoids his hour-and-a-half commute into Washington, D.C., altogether by working at home. "I can do just about anything from my study that I can do from here," he said while visiting the office. And, he added, "I get to be home for dinner."[30] Avoiding a time-consuming and often frustrating daily commute is a major impetus to many of those who support telecommuting.

CHANGES IN OUR SOCIAL INSTITUTIONS

The impact of the Industrial Revolution extended well beyond organizations and work itself to our social institutions.[31] Perhaps the most notable are changes to the family and the role of women. Most U.S. colleges and universities began accepting women in the mid-1900s. By 1995, 55% of all bachelor's degrees earned were granted to women.[32] In addition to larger numbers of women in colleges and universities, today more women are in the workforce than ever before. Women have more career choices and opportunities than at any time the past. Although a wage gap still exists between men and women workers, that gap is shrinking. This is largely due to an increase in women's educational achievements and their ability to contribute in a knowledge economy at the same level as men.

Along with the many benefits of increased numbers of women in the workplace comes the increased challenge of achieving life balance. Balancing work and family means having time to cook, clean, shop, launder, care for the children, perform well on the job, and still have some time to relax and enjoy life.

More women in the workforce is not the only factor adding to the difficulty of this challenge. Baby boomers, technological advances, and a good economy have led to increased opportunities and expectations. People today are busier than ever before, trying to squeeze as much as they can out of each day. Even our children have schedules that would boggle the minds of children a generation ago. We keep family calendars on the refrigerator so we can keep track of where everyone is and where everyone has to be. Even parents who don't work outside the home can have daunting schedules, completing chores and errands during the day and spending evenings transporting children to various appointments, lessons, and competitions. For many families, dinners together must be scheduled in advance or enjoyed from the interior of a minivan as it pulls away from the drive-through window. For many, the bustling leaves little time for reflection, but in those few moments we do have, we wonder if something important was left off the to-do list.

For many people seeking employment as telecommuters, the driving force behind telecommuting is the desire to find balance in work and family. For some, that comes with the flexibility of the work schedule and the time saved by not having to commute. Many telecommuters, both male and female, talk about the importance of the flexibility that telecommuting offers them. "It allows me to participate fully in the rearing of my children," says Dave Brady, a special project manager in computer chip design with InterGraph Electronics. "It allows me to work the hours which suit me."[33] Telecommuters can be there when their children come home from school; they can make it to the school play; and they can be in the stands for the big game.

Virtual Wisdom ▼

It's not a benefit like life insurance that everyone's got to have. It's a work option that both the manager and the employee come to an agreement on.
—Judy Rapp-Guadagnoli, consultant[34] ▲

ADVANTAGES OF VIRTUAL WORK ENVIRONMENTS FOR EMPLOYERS

Working from home is not always the result of a choice made by an employee. In some cases, offices are phased out, global organizations may not have regional office space, or companies may want to expand their staff without expanding their facilities. Start-up companies may not have the capital for real estate and may not need it if they can operate as a completely virtual organization.

Organizations see the benefit of having less office space and parking area to support. According to Jack Heacock, member of the executive committee of the International Telework Association & Council, "At 10:45 a.m. on any particular day, 40% to 60% of employees are somewhere other than their offices. Walk around and see for yourself. Then ask yourself why you are investing in real estate space you don't need."[35] Office space sits empty in part because employees are traveling or working at client locations. When these employees travel, their work travels with them. They use cell phones and laptops to stay connected and make their office wherever they happen to be for the day. Many organizations now recognize that maintaining empty desks is an expense that can be avoided, since employees are able to work from nearly anywhere. Organizations wanting to cut overhead costs can reduce their real estate expenses by having employees telecommute, if feasible. Northern Telecom estimates that it saves about $2,000 per year, per person for each office eliminated. IBM has reduced its annual real estate expenses by as much as 40% to 60% at those locations where all offices (except those deemed essential) are eliminated.[36]

Another reason for organizations to consider employing telecommuters has been the demand for highly skilled technical experts. In the tight labor market of the 1990s, hiring and retaining the best employees was a genuine challenge. Sun Microsystems opened three telework centers around the San Francisco Bay area to appeal to its employees. It chose the location of the centers by demographics. "Instead of chasing space, we're chasing people," said Ann Bamesberger, director of workplace effectiveness. "Instead of thinking, 'Where can we get a big chunk of land?' we're thinking, 'Where do people want to work?'" The centers have been so successful that the company plans to open more.[37]

Because telecommuters can work from anywhere, allowing telecommuting significantly enlarges the labor pool. Companies now have greater access to technical experts and a larger selection of potential job candidates than ever before. Many organizations recognize that the most desirable group of people for any one job is often distributed across numerous locations. Creating a virtual team allows organizations to access that talent without the expense of relocating the employees, if they can be convinced to move.

Virtual teams offer a number of advantages. Through virtual teams, professional communities that share knowledge can develop from separate islands of knowledge. In this way, virtual teams can develop and spread best practices faster across organizations, as they foster cross-functional and cross-organizational collaboration. Virtual teams enable companies to respond faster and at lower cost to market changes, global competition, and shortened product life cycles. Companies can focus on their core competencies by forming virtual partnerships with other companies in order to outsource non-core needs.[39] Managers recognize and appreciate a

Virtual Wisdom ▼

It's the skills and talent that are important, rather than face time.

—Tom Vines, Director of Talent, IBM[38] ▲

Virtual Wisdom ▼

We are part of a global company with operations in 100 countries and over 500 locations. We wanted to capitalize on talent within our organization and we made a commitment to use technology and e-business practices to do so.

—Joy Gaetano, Senior Vice President of Human Resources, USFilter[40] ▲

reduction in cost afforded by virtual teams, and team members see an advantage in increased independence and greater flexibility.[41]

Telecommuting has also been shown to save money for organizations through reduced absenteeism and increased retention of employees. The same flexibility that allows employees to take a child to the doctor or take a car in for repairs during regular working hours also allows that employee to make up for those hours during other times of the day or night. Research suggests that the savings to organizations is around $2,000 per employee in reduced absenteeism and more than $7,000 in reduced turnover costs.[42] Merrill Lynch implemented telecommuting in 1996 and today has over 3,500 employees that work from home one to four days a week. Managers at Merrill Lynch report that they saw 3.5 fewer sick days per year and a 6% decrease in turnover among their telecommuters during the first year of the program.[43]

Of course, companies concerned about the bottom line want not only to reduce overhead but to simultaneously increase productivity as well. Merrill Lynch saw a 10% to 50% increase in productivity with their telecommuters.[44] Internal research by organizations that have implemented telecommuting programs, including IBM, AT&T, and American Express, reports productivity gains from their telecommuting.[45] "It increases productivity enormously," said Bill Finkelstein, a senior partner at Cisco Systems. About half of Cisco's 35,000 employees work from home at least one day per week. Finkelstein notes that, for himself, eliminating time in the car means he can start his workday at 6 a.m.[46] Telecommuters themselves say that they are as productive as or more productive at home than they are in the office. Some attribute their increased productivity to having fewer interruptions at home. They spend less time in meetings, for example, than they would in a traditional office.

Other ways that telecommuting can save companies money include

- Saving on employee relocation costs,
- Paying lower wages in other countries, and
- Lowering travel expenses by locating employees closer to customers.

In addition to benefits to the bottom line, telecommuting has implications for the environmentally and socially conscious organization. Telecommuting benefits the environment. Automobiles produce six of the seven pollutants for which the Environmental Protection Agency has created health standards. These pollutants include carbon monoxide, nitrogen oxides, lead, volatile organic compounds, sulfur dioxide, and particulate matter. Automobiles are the number-one producers of carbon monoxide and nitrogen oxides.[47] One government study found that carbon emissions could be reduced by 81,600 pounds per week if 20,000 federal employees telecommuted just one day each week. In addition, they would save a cumulative total of 102,000 gallons of gas.[48]

Socially conscious organizations, by offering a work-from-home option, can help remove barriers to employment experienced by some individuals. People with physical disabilities that prevent them from traveling to a job have been separated from gainful employment by transportation barriers—barriers that are easily overcome by telecommuting. Location can in itself be a barrier to gainful employment. In the United States, pockets of poverty have persisted in rural areas where the land is not well suited to farming and there is little industry. For people who live in such areas and are not able to move, virtual work environments can provide them with a means for employment and future opportunity.

WHY AREN'T WE TELECOMMUTING MORE?

With these numerous incentives for telecommuting, you might well ask why we're not doing more of it. Forecasters at one time predicted that 55 million U.S. workers would be telecommuting by the early 2000s, far above the 23 million that are estimated to have telecommuted last year. Surveys show that more than 40% of workers report that their jobs could be performed, either entirely or in part, from a remote location if they had the appropriate communication devices available. However, only about one-fifth of the workers said that their employers actually allowed them the opportunity to work off-site.[49] Dawn Silvia, a Boston public relations consultant who has been searching for a telecommuting job for four months, complains that when it comes to telecommuting, "a lot of companies talk the talk, but they don't walk the walk." She has received offers from companies that would like very much to hire her—as long as she doesn't work from home. "They're all saying the same thing: 'We want you in here, at a desk where we can watch you and trust that you're doing your job.' "[50]

The most significant obstacle for telecommuting seems to be management reluctance. Managers who have learned and practiced a line-of-sight management style are uncomfortable with the idea of having employees off-site. "A lot of us who learned our management techniques or styles in the 60s or 70s have sort of a bias that we have to see our employees at work," asserts Paul Chistolini, acting commissioner of the public buildings service of the U.S. General Services Administration. "It's quite a culture change to have to let people go and simply trust them to do the work."[51] The bias toward managing by seeing hasn't changed much, and many managers are unsure of how to train and monitor the performance of remote employees.

Middle managers also have concerns about becoming irrelevant. Remote workers are more self-directed and the hierarchies within organizations with telecommuters are, by necessity, flatter. Middle managers are frequently concerned about their perceived value, especially when all of their direct reports are out of the office and, thus, out of sight. Middle managers may be concerned that upper management will see the empty office space and wonder whether the middle manager is really needed.

Virtual Wisdom ▼

The command-and-control management mentality really gets less and less successful.
    —William Bridges, founder
    William Bridges and Associates[52] ▲

There are several other factors that inhibit organizations from instituting telecommuting programs.

- **Set-up costs.** The initial outlay to purchase equipment can be significant. In order to be effective, off-site workers must have the right communication tools and technologies that are compatible with those in the rest of the organization. At the very minimum, telecommuters will need a computer, an all-in-one printer that will scan, copy, and fax, a separate phone line for work, and office furniture. The cost of equipping a remote worksite varies from $3,000 to $5,000.[53]
- **Loss of cost efficiencies.** When everyone works from a centralized location, all of the most expensive equipment can be shared. Not so with off-site workers.

- **Technical challenges.** Managers have concerns about providing technical support to remote workers. If the printer stops working in the central office, you call IT maintenance, but what do you do if the printer stops working hundreds, or even thousands, of miles away?
- **Culture.** In many cases, the organizational culture has not been recrafted to appreciate telecommuting. The belief persists that working from home is not really working. Some managers are concerned that telecommuting creates resentment in office-bound workers and that this will have a negative affect on organizational culture and company loyalty.[54]
- **Security.** Some managers are concerned about the security of sensitive data used at a remote location and are overwhelmed when they read all the high-tech information available on securing networks.
- **Tax issues.** Many managers are uncertain of the tax and legal issues surrounding telecommuting. According to a study conducted by the U.S. General Accounting Office (GAO) for the purpose of identifying barriers to telecommuting, organizations are uncertain about the application of state laws to telecommuting. They worry that telecommuting may expose organizations to additional corporate taxes of employees or to additional state taxes if remote workers are located in another state.[55]
- **Safety issues.** Also identified as a barrier to telecommuting by the GAO are concerns about employee safety. Currently, the Occupational Safety and Health Administration (OSHA) does not require an employer to assume liability for the safety of a home office. Some organizations are concerned that that may change.[56]
- **Intercultural clash.** Members of global virtual teams will have to work through national cultural differences to succeed. Even the differences in organizational cultures and business practices of multi-organizational teams may present a challenge for team members.
- **Communication issues.** When people don't regularly see each other face to face, many of the communication problems that are encountered when people work together are exacerbated. For example, it's much easier to misunderstand a person's meaning or intentions without nonverbal cues such as facial expression or tone of voice to help with interpreting a message. This is especially true for global virtual teams where the members don't share the same culture.
- **Building trust.** One important challenge for virtual organizations is building trust. Trust is an essential element to working collaboratively with others and is usually built over time with the help of nonverbal cues. It is much more difficult to build trust among people that have little or no face-to-face contact and little non-task-related communication, such as within virtual teams, where research shows communication tends to be focused on the task at hand and relationships don't develop.[57]

In some cases, meager numbers of telecommuters aren't due to a lack of opportunities presented by the organization. Many larger companies that offer telecommuting report that only between 1% and 5% of employees participate in the programs.[58] A number of factors contribute to the choice by employees not to telecommute or to return to the office after telecommuting for a while.

- **Physical workspace.** Having a family member working at home offers some advantages for families but presents some challenges as well. One is very simply the workspace itself. People who work from home need a place to work, store their materials, and stay organized. In most cases, this is a section of a room that is used for some other purpose in the home. Even if a separate office exists in the home, this space is very often shared with other members of the family.

- **Psychological workspace.** Finding the physical space is only half the problem. Protecting the psychological space is the other half. Even for those telecommuters fortunate enough to have an office set up in a separate room, the presence of other family members in the house (such as children home for summer break) can make it difficult to find quiet blocks of time to work. Silvia Orian, a sales representative for Xerox, has been telecommuting for two years. "The most difficult thing for me is getting my family to realize that I'm at work," she said. When she really wants her family to know that she doesn't want to be disturbed, she will close the door to her office, and this seems to help.[60] People working at home may be distracted by their families, even when the family isn't being particularly intrusive. Men tend to see the family as a temptation; they want to play with the kids. Women tend to see the family as a responsibility.[61] This can be stressful for the female telecommuter, who may experience role conflict when she is working in the home and feeling as if she should be caring for the family instead.

- **Learning to log off.** Those who work from home may take time out of the workday for family, but they pay that time back. The flexibility of the work schedule that many seek through telecommuting may lead to a blurring of the distinction between work and non-work time. With communication technologies such as pagers, cell phones, and personal digital assistants (PDAs), telecommuters can be reached 24 hours a day. Many people already feel that this constant accessibility is an intrusion, and, at some point in the future, they may rebel against always being available. For many teleworkers who experience the lack of boundary between work and non-work modes, however, it isn't the others in the virtual organization or team who are the problem, but the teleworker. "The problem I have is disciplining myself to get out of the office," said Bill Finkelstein of Cisco about working from his home office.[63] It can be difficult for a telecommuter to disengage if there is work right in the next room to do. Despite the desire to balance work and family, teleworkers can begin to feel as if there is never a real day off from work. Teleworkers don't ever walk away from the office and leave the workday behind, so it is important that they discipline themselves to log off now and then.

> **Virtual Wisdom ▼**
>
> *I have an office on the third floor that's separate from our living area, and if I want to separate myself, I just close the door.*
>
> —Jonathan O'Keefe, telecommuter
> Global Technology Services[59] ▲

> **Virtual Wisdom ▼**
>
> *If I want to truly take a day off, I have to leave the house.*
>
> —Craig Meadows, telecommuter[62] ▲

It's easy to be overlooked when promotions are being handed out if no one can see you.

—Nick Hough, President, Australia Pacific
Teleworking Association[64] ▲

- **Remote workers' concerns about career advancement.** Managers are not the only ones concerned about the possible out-of-sight, out-of-mind effect. Remote workers see challenges in recognition, isolation, receiving technical support when needed, and communication issues. Some experts have suggested that working off-site can lower one's visibility to a point where career growth may be stunted.[65]
- **Perceived legitimacy of having an office.** There is something about having an office that adds a feeling of legitimacy to any business venture. The person that starts a business out of the family garage or basement appears legitimate and more likely to succeed when he or she finally gets an office.[66]

Many of the barriers to success in a virtual organization are communication-based. Some are more obviously so than others. For example, it's clear that building trust and building relationships on virtual teams are communication issues. It may not be as clear that the perceived legitimacy of having an office is a communication issue, as well. Having an office, the location of the office, the size of the office, and the appearance of the office all communicate information (e.g., professionalism, conservatism) about an organization to its employees and clients. In a virtual workplace, that information must be communicated in other ways, such as in the appearance of the company's web page or the promptness with which messages are returned.

In the remainder of this book, we will talk about the special challenges faced by those who work in virtual environments and those that manage them. In the following chapters, we will describe the communication process and how it is affected by a virtual context. We will do a brief survey of the communication appliances available to virtual workers, discuss media choices for communicating in a virtual workplace, and explore management issues that relate to communication barriers in the virtual workplace.

DISCUSSION QUESTIONS

1. Consider your current or previous job. What portion of that job could be performed away from the office? Was telecommuting an option available to you in this position? Was it ever discussed?

2. Describe how technological advancements have affected family life, friendships, work, and school. In what ways are these effects positive and in what way are they negative?

3. What do you think would be the most significant advantage of being an employee in a virtual organization where telecommuting was full-time? What would be the most significant challenges? If you were to telecommute, would you prefer to do this every day of the week, or only some days of the week?

4. How could telecommuting be used to address the problems of inner-city unemployment and the technology gap? Why do you think organizations would or would not develop a program to address these issues?

5. What do you think are the most significant challenges for managers of virtual organizations? What management skills do you think are most important in the virtual workplace? What style of management do you think is the most effective?

6. How do you think that an organization could use telecommuting to improve its corporate reputation?

ENDNOTES

1. Elizabeth Allen, personal communication by phone with author, June 24, 2002.
2. Anthony M. Townsend, Samuel M. DeMarie, and Anthony R. Hendrickson, "Virtual Teams: Technology and the Workplace of the Future," *Academy of Management Executive,* vol. 12, no. 3 (1998): 17–29.
3. Michael A. Verespej, "The Old Workforce Won't Work," *IndustryWeek.com,* September 1998.
4. Deborah L. Duarte and Nancy T. Snyder, *Mastering Virtual Teams* (San Francisco: Jossey-Bass Publishers, 1999).
5. Ibid.
6. Silvia Orian, personal communication by phone with author, June 20, 2002.
7. W. W. Cooper and M. L. Muench, "Virtual Organizations: Practice and Literature," *Journal of Organizational Computing & Electronic Commerce,* vol. 10 (2000): 189–209.
8. Jonathan W. Palmer and Cheri Speier, "A Typology of Virtual Organizations: An Empirical Study," The University of Oklahoma. Available: *http://hsb.baylor.edu/ramsower/ais.ac.97/papers/palm_spe.htm.*
9. Martha Haywood, *Managing Virtual Teams: Practical Techniques for High-Technology Project Managers* (Boston: Artech House, 1998).
10. John Gunn, "The Virtual Corporation," December 2001. Available: *http://bcn.boulder.co.us/business/BCBR/december/virtual.dec.html.*
11. Palmer and Speier.
12. Dana G. Mead, "Retooling for the Cyber Age," *CEO Series* 42, September 2000. Available: *http://csab.wustl.edu/csab/.*
13. Katie Hafner, "Working at Home Today?" *New York Times,* 2 November 2001, D1.
14. Gil Chiquito, personal communication with author, June 12, 2002.
15. Kevin Voigt, "Some Enterprising Employees Exchange the Cubicle Life for Pools in Exotic Places," *Wall Street Journal,* 31 January 2001, B1.
16. Carl E. Van Horn and Duke Storen, "Telework: Coming of Age? Evaluating the Potential Benefits of Telework," *Telework and the New Workplace of the 21st Century,* 2000. Available: *http://www.dol.gov/asp/telework/p1_1.htm.*
17. Ibid.
18. Ibid.
19. Glenn Lovelace, "The Nuts and Bolts of Telework: Growth in Telework," *Telework and the New Workplace of the 21st Century,* 2000. Available: *http://www.dol.gov/asp/telework/p1_2.htm.*
20. Jonathan Glater, "Telecommuting's Big Experiment: Federal Employees Are Urged to Work Outside of the Office," *New York Times,* 9 May 2001, C1.
21. Alex Thio, *Sociology,* 5th ed. (New York: Addison Wesley Longman, 1998).
22. J. Russo and P. Schoemaker, *Decision Traps* (New York: Doubleday, 1989).

23. Paul Recer, "Home Computers, Web Access Grows in U.S.," *The Detroit News,* June 20, 2000, Available: *http://detroitnews.com/2000/technology/0006/20/a10-78208.htm.*
24. Thio.
25. Katie Hafner, "Company Sites Are Holding Offices Together," *New York Times,* 26 September 2001, D1, D8.
26. Max Weber, *From Max Weber: Essays in Sociology.* Translated and edited by H. H. Gerth and C. Wright Mills (New York: Oxford University Press, 1946).
27. Verespej.
28. Mead.
29. Hafner.
30. Glater.
31. Thio.
32. Ibid.
33. Caron Schwartz Ellis, "Telecommuting Quickly Becoming Benefit for Employer, Employee," *Boulder County Business Report,* 1995. Available: *http://bcn.boulder.co.us/business/BCBR/1995/sep/commute2.html.*
34. Ibid.
35. Michael A. Verespej, "The Compelling Case for Telework," *Industry Week,* vol. 250 (September 2001): 23.
36. Wayne Cascio, "Managing a Virtual Workplace," *Academy of Management Executive,* vol. 13, no. 3 (2000): 81–90.
37. Hafner.
38. Carla Joinson, "Managing Virtual Teams," *HR Magazine* (June 2002): 69–73.
39. Haywood.
40. Joinson.
41. S. Eom and C. Lee, "Virtual Teams: An Information Age Opportunity for Mobilizing Hidden Manpower," *S.A.M. Advanced Management Journal,* vol. 64 (1999): 12–17.
42. Susan J. Wells, "Making Telecommuting Work," *HR Magazine,* vol. 46 (2001): 34–46.
43. Ibid.
44. Lovelace.
45. Ibid.
46. Hafner.
47. "Road Trouble Ahead," 2001. Available: *http://www.cleanair.org/green/roadtroubleahead.html.*
48. "Mobile Source Air Toxic Emissions," 2002. Available: *http://www.epa.gov/otaq/toxics.htm.*
49. Van Horn and Storen.
50. Kemba J. Dunham, "Telecommuter's Lament," *Wall Street Journal,* 31 October 2000, B1, B18.
51. Glater.
52. Verespej, 1998.
53. Cascio.
54. Dunham.
55. United States General Accounting Office, "Telecommuting: Overview of Potential Barriers Facing Employers," July 2001. Available: *http://www.gao.gov/new.items/d01926.pdf.*
56. Ibid.
57. S. Jarvenpaa and D. Leidner, "Communication and Trust in Global Virtual Teams," *Organization Science: A Journal of the Institute of Management Sciences,* vol. 10 (1999): 791–846.
58. Wells.
59. Nick Grabbe, "Telecommuting Works for Many," February 7, 2000. Available: *http://www.gazettenet.com/biz2000/02072000/21636.htm.*
60. Orian.

61. Kiran Mirchandani, "Legitimizing Work: Telework and the Gendered Reification of the Work/Nonwork Dichotomy," *Canadian Review of Sociology and Anthropology,* vol. 36 (1999): 87–107.
62. Grabbe.
63. Hafner.
64. Elizabeth Walton, "Virtual Office." Available: *http://www.pnc.com.au/~lizzi/Virtual_office.html.*
65. Wells.
66. Van Horn and Storen.

2 THE COMMUNICATION PROCESS

Communication is often thought of as the transfer of information from one person to another. However, this definition makes communication sound like an event, rather than a process. True communication refers to the process by which shared meanings are created. You make a statement to a colleague who gives you a confused look. You restate your point in another way and your colleague nods in understanding and paraphrases what you said. You respond with an affirming, "Right."

The process of communication is fundamental to nearly all organizational activities. But effective communication is often difficult. In surveys of managers, communication repeatedly ranks high as an area of concern.[1] Communicating effectively is even more challenging in virtual work environments. In organizations, effective communication is evident by the achievement of specific communication objectives. The challenge for virtual workers is to achieve these objectives without the regular use of face-to-face interactions and the accompanying nonverbal cues.

A model of the communication process (see Figure 2-1) begins with a stimulus, something that causes the sender to want to communicate. The sender then encodes a message. The sender transmits the message after selecting a channel (verbal, nonverbal, spoken, or written) and then a medium (e.g., phone call, e-mail, face-to-face meeting). The receiver decodes the message and responds with some sort of feedback.[2] A change in any of these interconnected elements will have an effect on the rest of them. If you, as a sender of a message, are in a wonderful mood, that will affect your communication with others. If the integrity of a message delivered by telephone is compromised by static interference, the entire interaction is affected.

Communication is a process that, once begun, continues and is always changing. The process of communication is not only continuous, it is irreversible and unrepeatable. Once communication occurs, it can never be undone and can never be repeated as new.

The decisions that communicators make as they move through the process will affect the outcomes. For example, suppose the communication objective for a human resource manager is to inform employees of an unavoidable, negative change to their benefits. If the only objective is to impart the information, the manager has a number of equally effective options: send a memo or an e-mail simply stating the change. But if the real objective is to impart the information in a way that prevents employees from reacting angrily and without negatively affecting morale and productivity, then she must be more thoughtful in choosing a medium and composing her message. She must consider how her audience will respond to her different media

Figure 2-1   Model of Communication

Human communication is a process in which meaning is transferred as senders encode messages to be transmitted over some medium to be decoded by receivers. All of this takes place dynamically and continuously in the presence of noise, against the background of the participants' experience, is governed by a set of ethics, and produces some effect.

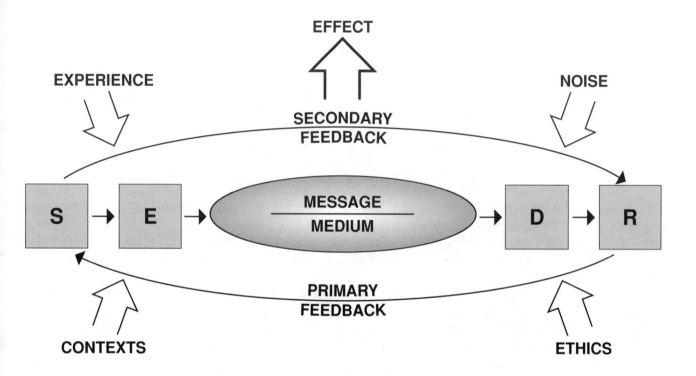

choices. Would voice mail convey more caring than a memo? When she crafts her message, she will want to use a tone that conveys empathy, and she should include the reasons for the change so that the employees know it wasn't made arbitrarily without considering their needs. Making the right choices as she moves through the process will improve her chances of reaching her communication objective. Whether in a traditional or virtual work environment, concerns for audience, purpose, and occasion remain paramount in effective communication.

COMMUNICATION IN CONTEXT

The communication process takes place in a physical context. The physical context includes the location of an interaction and the physical aspects of that space. We can easily understand the effect of physical context on the process of communication by considering the different forms the same piece of information might take when delivered in a boardroom as opposed to on the golf course. The very same underlying message may be formed differently or interpreted differently in dissimilar physical surroundings. In the boardroom, for example, the message may be delivered using more formal language and different nonverbal cues than on the golf course.

Like the physical context, the psychological context can have an affect on the communication process, as well. An example of psychological context might be the communication culture of an organization. Some organizations are very open and encourage free and frequent communication. Other organizations are more closed and promote a feeling that communication is not often welcome, fostering less communication. Another example of psychological context and its impact might be the relationship between the sender and the receiver. The same message may take different forms and be delivered using different media, depending on their relationship. For example, an employee's suggestion for a procedural change may be delivered in casual conversation to a co-worker or friend but polished up in a formal memo to the boss.

The virtual workplace is an interesting communication context, because the sender and receiver do not share a physical context for communication. In many cases, they don't share a psychological context, either. For example, when high-ranking executives visit a facility, a nervous energy spreads through the organization. Remote workers are disconnected from this experience, unless an effort is made to at least make them aware of it. A message sent from the home office to a remote worker may not directly state this feeling but the worker may still be affected by it in some way.

CHALLENGES TO EFFECTIVE COMMUNICATION

Missing from the model thus far is anything that interferes with the perfect transfer of information. As communicators, we all know that effective communication is neither perfect nor easy.

James O'Rourke, in his book *Management Communication*, notes that the act of communicating is not the same thing as the process of communication. Simply sending a message doesn't mean that you have effectively communicated.[3] Anything that disrupts the communication process and interferes with the creation of shared meaning is referred to as noise.

Virtual Wisdom ▼

Candidates who come off poorly in telephone interviews or in e-mail exchanges probably aren't going to make great remote workers.

—William Pape, cofounder
VeriFone, Inc.[4] ▲

NOISE

Noise, like context, can be physical or psychological and can be associated with any of the elements of the model of communication. Physical noise, for example, could be office music that is too loud, people talking during a presentation, or the sounds of a road crew at work outside an office while you're trying to talk on the phone. Psychological noise refers to internal sources of distraction, such as a receiver's preoccupation with something else, a receiver's feelings about the credibility of a sender, a sender's prejudgment of a receiver, a message that contains emotionally

Table 2-1   Examples of Psychological Noise

Sender Noise	Receiver Noise	Message Noise
Poor delivery skills	Stereotypes of the speaker	Poorly constructed message
Preconceptions about the listener	Lack of motivation	Extreme message
Communication apprehension	Missing background information	Complex message

arousing words, or a medium that violates the receiver's expectations for message delivery. Table 2-1 provides examples of psychological noise from three points of view.

SYNCHRONICITY AND FEEDBACK

A quick examination of the model of communication would make it seem as if communication occurs in a continuous loop of turn-taking between sender and receiver. But as communicators we all know that, at least during face-to-face communication, receivers of our messages give us feedback *while* we are sending our message. If our receiver is bored, excited, or hurt by what we are saying, it is unlikely that we will have to wait until it is the receiver's "turn" in order to receive feedback about that reaction. A glance at the clock, a frown, a tear—many sorts of feedback can occur while we are sending a message. Synchronous communication occurs at the same time and is, therefore, simultaneously interactive. While sending a message, we are also picking up on the feedback to that message.

This synchronous communication is limited in a virtual workplace. Face-to-face interaction is the exception rather than the rule for communicating. Asynchronous communication, which doesn't occur at the same time, can add to the challenge of effective communication for both the senders and receivers of messages. Receivers must try to understand messages without the assistance of nonverbal cues, and senders must determine the receiver's understanding of messages without immediate feedback.

In a traditional organization, much communication is written, but the primary means of day-to-day communication is face to face. In a virtual organization, the primary means of communication is likely to be written in the form of e-mail. A summary of some of the differences between written and spoken messages is offered in Table 2-2.

COMMUNICATION IN ORGANIZATIONS

Research on communication in organizations has frequently focused on how communication flows in different organizational structures. Formal communication networks in organizations can be centralized, where information flows to and from one central person, or decentralized, where each member in an organization has the same access to information, or somewhere in between these two extremes.

Table 2-2　　Written Communication versus Face-to-Face Communication

Written Communication (Asynchronous)	Face-to-Face Communication (Synchronous)
Is a persistent, static "object"	Is ephemeral
Can be reread anytime	Cannot ordinarily be reviewed
Can take time before responding	Must respond immediately while in the interaction
Consists of discrete separate symbols	Is virtually continuous
Consists of easily separated words	Consists of words and other acts that merge
Requires no audience present	Requires the audience to be present
Means the writer and the reader may not share a setting	Means the participants are in the same setting
Is learned at school	Is learned at home
Is taught with explicit, conscious norms	Is practiced rather than explicitly taught

Sources: Janet Beavin Bavelas and Nicole Chovil, "Visible Acts of Meaning," *Journal of Language and Social Psychology,* vol. 19 (2000): 163–195. Per Linell, "The Written Language Bias in Linguistics" (1982). Available: *http://eserver.or/langs/linell.*

Information in organizations can travel up, down, or across the organizational hierarchy. Downward communication travels from upper management down and includes directives, instructions, performance evaluations and feedback, and the mission, vision, and strategy of the organization. Upward communication travels from the lower levels up and includes information such as problems, performance reports, accounting information, and issues not related to the work itself, such as requests for time off. Horizontal communication refers to messages that travel between employees at the same level in an organizational hierarchy.

Virtual organizations are recognized as essentially possible only with decentralized, nonhierarchical networks. However, research on the communication patterns that emerge in virtual organizations suggests that centralized networks may be preferred when they are more appropriate for the task at hand, such as with a routine task. The issue appears to be one of a fit between task and network structure. The important implication for managers and leaders of virtual teams is to not assume that a nonhierarchical communication structure is automatically the best choice for every situation. Monitoring and managing communication structures by task is a better strategy.[5]

Virtual Wisdom ▼

The key to making it work is simply "knowing who you're communicating with and messaging about the right things."

—Mark Gibbs, Network World[6] ▲

Organizations have informal communication networks called grapevines. Information traveling along the grapevine can circulate through an organization very rapidly. Grapevines can serve a valuable social function in organizations and give managers a source to tap into information that may not be circulated upward and to leak information downward. But grapevines have their dark side. Rumors are often circulated through grapevines to the detriment of an organization and individuals within it.

Work groups that communicate primarily by e-mail, as in virtual organizations and teams, appear to have less informal, non-task-related communication.[7] While avoiding the rumor mill may be a positive aspect of having fewer informal networks, it is more than offset by losing the benefits, particularly the social function. Off-site workers and those that manage them should encourage the creation of informal networks and non-task-related communication to build relationships.

COMMUNICATION BARRIERS IN ORGANIZATIONS[8]

Barriers to communication exist in all communication contexts. Organizations are prone to any one or more of the following barriers.

- **Differences in perception.** An important type of psychological noise occurs when there is a difference between the sender and receiver in the perceived meaning of a message. Perception is the mental process of interpreting and assigning meaning to the sensory information that we receive through our senses. How we perceive or interpret sensory information depends on a number of factors, including culture, personality, experience, and mood. Therefore, individual perceptions of the same sensory information can vary tremendously. Jurors who hear the same evidence, for example, will often have diverse views of what the evidence means in terms of the defendant's guilt or innocence. Communication is complicated by the fact that we each have our own perception or interpretation of a message. The challenge for us as senders and receivers is to make those interpretations similar enough to accomplish our communication objective.
- **Information overload.** As Wayne Cascio, Professor of Management at the University of Colorado, says, "You can't look at everything."[9] Most of us start our workday with a flood of e-mails, voice mails, memos, and snail mails. In addition to the torrent of incoming messaging, we have access to enormous amounts of additional information. Technology has opened the information floodgates and some people are drowning as a result. With so much information available, it's hard to know which information to pay attention to and when you have enough. When sorting through incoming messages, people suffering from information overload may ignore messages or process only part of a message. For example, they may read only the subject line of a memo or the first line of a letter. If they are not convinced at that point that the message is important to them, they will discard it without reading the rest. When wading through available information, they make hurried and often poor decisions. Because they spend so much time amassing information, they are unable to spend adequate time reviewing and evaluating it.

  Organizations with inefficient communication practices encourage the problem of information overload. E-mails may be sent to broad lists even though the messages don't apply to many of those on the list, or meetings may be held where attendance is required, even for attendees that will gain or contribute little.

Often telecommuters will cite avoiding these sorts of distractions at work as a reason for choosing to work off-site. However, information overload is still a problem for those in virtual work environments and can be a damaging one. For on-site employees, serious messages are often delivered face to face. Even when a message is delivered by another medium, a casual encounter in a hall gives a manager an opportunity to confirm that an employee received and understood a message. If an employee erroneously considered a memo unimportant based on a glance at the subject line, an exchange at the coffeepot could lead him to rethink that conclusion and retrieve the memo from the trash. In other words, informal communication in the organization provides a failsafe mechanism for on-site employees that remote employees do not have. Managers can address this issue by putting formal feedback structures in place for the receipt of important messages.

- **Message complexity.** Today most of us can easily relate to experiencing difficulty when trying to understand very technical information. It can be just as difficult to prepare a message about a complex topic in a way that is clear. In a virtual work environment, where e-mail is the primary means of communication, immediate feedback, such as a clarifying question, a nod of understanding, or a completely confused look, is not available while the message is being sent. Both senders and receivers in virtual work environments must strive for clear messages and feedback.

- **Differing status.** People occupying different levels on the organizational hierarchy may have difficulty communicating with each other for various reasons. Managers may not adequately value the knowledge of lower-status employees, and lower-status employees may resist sharing negative information with a manager because of concerns about the impact on future promotions. This barrier may be less of a concern for those working in virtual organizations than for those in traditional ones. Differences in status are not as prominent in virtual environments, where larger offices and nicer furniture

are unseen. Furthermore, people tend to be more willing to report negative information when they are able to do it through e-mail rather than face to face.

- **Unethical communication.** In traditional and virtual organizations, a company that practices unethical communication will not be trusted by its employees. Distorted, incomplete, or deceptive information damages internal and external relationships and destroys credibility.

- **Physical distractions.** Remote workers, especially if they are telecommuting from home, may have special concerns about physical distractions. Uncomfortable home office furnishings, inferior equipment, and household noises can all detract from the concentration of the telecommuter.

- **Incorrect medium choice.** While this is a concern for people in traditional, as well as virtual, offices, remote workers are faced with additional choices of advanced communication technologies when selecting a medium. This topic is discussed at length in Chapter 4.
- **Lack of trust.** In environments without trust, people do not communicate openly. Building trust is an important and recurring issue for virtual organizations and teams because much of what leads people to trust others is communicated nonverbally. We will discuss communicating trustworthiness in Chapter 4.

NONVERBAL COMMUNICATION

In traditional organizations, the spoken word is the most common form of communication, while the written word, in the form of e-mail, is most common in virtual work environments. What are the implications of this difference for the communicator? One is the limited nonverbal cues available through e-mail. Nonverbal communication can provide cues to the interpretation of verbal messages, as well as communicate independently of them. Your verbal message is about 7% of your communication; your tone of voice accounts for about 38%, while your body language delivers around 55% of your communication.[11]

When people communicate face to face, they perform a symphony of nonverbal behaviors. In a study on the richness of everyday face-to-face dialogue, researchers found that not only do speakers emphasize, repeat, or replace words with nonverbal cues, they also provide images to accompany their messages. Very often, facial expression, gestures, and tone of voice may communicate different aspects of a message as a whole. For example, one of the participants in the authors' study described her brother's experience in a train station when he was unsure of what train to take. She made the statement, "He's trying to decide which one to take, ya know." When making this statement, she waved her parallel hands from right to left suggesting the positions of the trains he was choosing from, while at the same time widening her eyes to portray her brother. While she spoke, she was smiling at the listener. So with the delivery of the brief statement, the speaker simultaneously created a scene with her hands and eyes, and connected with her listener through her smile, which was not part of the scene.[12]

Certain categories of nonverbal behavior, for example, gestures, posture, facial expression, eye contact, tone of voice, and physical appearance, are lost when e-mail is used. In some situations, such as resolving a conflict or solving a difficult business problem, other means of communication, such as teleconferencing or videoconferencing that allow for nonverbal communication, are better choices than e-mail.

GESTURES AND POSTURE

Gestures and posture are body movements that convey meaning. The difference between a gesture and posture is that posture involves the movement of the body as a whole, whereas a gesture is accomplished through the movement of just one part of the body.[13] Numerous categories of body movements exist, but two, illustrators and affect displays, are particularly important to virtual work environments.

Illustrators help to show others the meaning of our words. If you explain to a non-native English speaker what the word *zigzag* means, you will quite likely use your finger to draw a zigzagging line in the air. Some people rely heavily on illustrators to help them communicate clearly and are challenged by limits to using only verbal communication. Affect displays are movements and postures that communicate emotions. When your roommate or spouse has a bad day, you can probably tell this by the way he or she is sitting or standing and how he or she moves. The same is true for your colleagues—unless, of course, they or you work off-site. The emotion behind a message may not be clear from its verbal content.

FACIAL EXPRESSIONS

The human face is capable of producing nearly 20,000 different expressions. It is unquestionably the most expressive part of the body. Most facial expressions are referred to as *facial blends,* meaning that they are a combination of two or more basic expressions. Many of our facial expressions are culturally learned, but some facial expressions (fear, anger, happiness, surprise, disgust, and sadness) have been identified as universally recognizable.[14] Facial expressions communicate the emotion of the message. Smiling as we attempt to communicate to those from another culture can help smooth the interaction. Without facial expressions, emotions can be misread. Smiling at a co-worker while making a teasing comment indicates that the comment is in fun. This same comment may be misinterpreted as an insult when it is separated from the sender's smiling face.

EYE CONTACT

Although the norms for eye contact vary across cultures, eye contact, or the lack of it, is a powerful way to communicate. We are all quite familiar with the implicit rules of eye contact. For example, what do you do if you're in class and haven't done the reading for the day, the professor is asking a question, and you don't want to be called on? Students and instructors alike recognize the "don't call on me" averting of the eyes. We use eye contact to tell whether people are listening to us; and when we want to get someone's attention, we try to make eye contact with them. In addition, research findings on eye contact indicate that it can help us to maintain control in meetings, overcome stage fright when speaking in public, and put others at ease.[15]

Important for members of virtual teams and collaborators, eye contact has been linked to evaluations of trustworthiness and competence. It is critical in the establishment of credibility. Communicators who maintain eye contact are perceived as more credible than those who avert their eyes, and listeners who maintain eye contact are perceived as better listeners. Ineffective eye behaviors, such as excessive blinking or looking at notes for long stretches during presentations, have been shown to lower credibility.[16]

PHYSICAL APPEARANCE

In our culture we place a great deal of emphasis on our appearance. At work, the importance of physical appearance in impressing a potential employer and establishing credibility and the perception of competence has long been recognized. For years, people have attributed different personality characteristics to individuals based solely on their physical appearance. In the 1940s, an American physician named William Sheldon developed a theory about body types and temperament. He devised three classifications for body type: endomorphic, a plump body type; mesomorphic, a muscular body type; and ectomorphic, a tall, thin body type. According to Sheldon, each body type corresponds to a particular temperament. Endomorphs, he theorized, are relaxed and sociable; mesomorphs are energetic and courageous; ectomorphs are introverted and restrained. Research that controls for cultural stereotypes associated with body type (e.g., fat people are jolly) shows little support for Sheldon's theory.[17]

Despite the lack of empirical support from research, in our everyday lives we frequently use stereotypes about appearance to evaluate people and form impressions of them. Stereotypes are collections of beliefs about a social group and members of that social group. Stereotypes are not always negative and are not always completely false. For example, research indicates that people hold positive stereotypes about attractive people. They tend to assume that more-attractive people are also happier, wittier, more intelligent, more popular, and more successful than are less-attractive people. Research shows that attractive people actually tend to have better social skills and less social anxiety; however, they do not have higher levels of self-esteem or overall happiness.[18]

Stereotypes are the result of forming social categories in order to simplify our thought processes. Age, sex, and race are primary categories by which we automatically group people as soon as we see them. Later, if we receive more information about people, we begin to think of them as individuals.

The problem with stereotypes is that they are very often negative, false, and overapplied. People tend to stereotype others negatively and then not look for individuating information. Holding a negative attitude toward others based on group membership (e.g., race, age, weight) is called prejudice. From its Latin roots, it means literally "to judge before knowing."

Stereotyping and prejudice often lead to discrimination. Discrimination is action taken toward an individual because of that person's group membership, which is often a minority group. For some groups, particularly the disabled, telecommuting is a means for overcoming some stereotypes and prejudices. Working off-site makes their disabilities less outstanding and allows the focus to be redirected toward their abilities.

OTHER FORMS OF NONVERBAL BEHAVIOR

PARALANGUAGE

Vocal cues, or paralanguage, include everything to do with an oral message except for the verbal content. Paralanguage refers to the "packaging" elements of a verbal message. These elements include rate of speech, tone of voice, volume, pitch, and the like. Imagine how many ways you can deliver a simple sentence like "See you later" or "Here you go" to communicate different emotions. The pitch, rate, and volume of your voice can give the listener information about how the verbal content of your message should be interpreted. Consider the statement "I didn't ask him to come to the meeting." Varying the tone of voice to emphasize different words changes the meaning of the sentence.[19]

- "*I* didn't ask him to come to the meeting." This suggests that someone else invited him to the meeting.
- "I *didn't* ask him to come to the meeting." This implies that someone doubts your denial.
- "I didn't *ask* him to come to the meeting." This could suggest that you mentioned the meeting, but didn't ask the man to come. Or you could be saying sarcastically that you didn't ask him to come to the meeting—you *told* him to come.
- "I didn't ask *him* to come to the meeting." This implies that you did ask someone else.
- "I didn't ask him to come to the *meeting*." This suggests that you asked him to come to something other than the meeting.

Our vocal cues also communicate to others about our self-confidence, our expertise, our credibility, and our professionalism. For a variety of reasons, most people would conclude that someone who uses frequent fillers, such as "umm" or "uhh" when speaking, or who pauses often, seeming to search for the next word, is less confident, less knowledgeable, less credible, and less professional than someone who speaks smoothly.

Vocal cues, like other nonverbal cues, are meta-communicational, meaning that they give us information about the communication. They help us to know how to interpret a verbal message. With remote workers using primarily written communication, verbal messages are not accompanied by vocal cues, increasing the importance of word choice and typographical cues that elucidate the meaning of a message.

ARTIFACTS

Artifacts are human-made objects—our possessions, our tools, and all that we gather up and call our own. They, too, can communicate about us. The homes we live in, the cars we drive, the books we read, the music we listen to—all communicate about us. Imagine that detectives who have never met you before enter your home while you are out. What could they learn about you by looking at your possessions? Think about your office. What sort of impression do you suppose the cleaning person, who comes into the office building at night and who has never met any of the employees (except the manager who hired him or her), has about you and your co-workers?

When it comes to off-site workers, the cleaning person may well have no impression of them at all. He or she may not even know they exist. And while this is of questionable importance, the fact that employees in the traditional office may also fail to form impressions of remote workers is of real concern. Off-site workers and virtual teams are often not represented in organizations by artifacts. The items in your office communicate about you, but a telecommuter doesn't have an office to hold the awards and certificates, or the pictures of the kids and family dog. Managers of off-site teams can keep them present in the minds of on-site workers by creating a space to represent them and to hold their artifacts. For example, a bulletin board in a hallway or break room where pictures, postcards, or announcements may be displayed can serve as a reminder of remote workers.[20]

SPATIAL ARRANGEMENTS

The use of space and the ownership or occupancy of space communicate status and organizational culture. Other aspects of the physical environment, such as the arrangement of objects and even the colors of carpets, walls, and office furnishings can also help to create and communicate organizational culture. The challenge for managers of remote workers is to impart this culture to those who do not share the physical environment.

TOUCH

Touch is another important way of communicating nonverbally. In the workplace, differences in touch behaviors have been studied as they relate to status and gender. Individuals vary in their need and desire to touch and be touched. In general, people are more likely to touch in social settings, rather than in professional settings. However, the amount of touching that is deemed appropriate varies across cultures, making it a topic worthy of further study for global virtual teams who may meet occasionally.

The most common touching behavior in the workplace is the handshake. One student enrolled in a class on communicating in a virtual organization indicated that she would find the absence of the initial handshake to be the most difficult adjustment to make in a virtual work-place. Handshakes communicate through the cleanliness of nails, the amount of dampness, the time spent in contact, and the firmness and style of grip. We have probably all experienced the unpleasant bone-crushing handshake or the equally unpleasant limp handshake. Most people prefer a firm handshake accompanied by eye contact. This type of handshake communicates feelings of respect and friendliness.[21]

SILENCE

Silence is also a very powerful nonverbal communication tool. If you have ever been given the silent treatment by someone who is angry with you, you know that silence can communicate loud and clear. Silence can communicate things other than anger, such as respect, relational comfort, or lack of interest. In a traditional organization, we are able to use context and other nonverbal cues to interpret the meaning of silence. For remote workers and those that manage them, silence can refer to unopened e-mails and unanswered phone messages. These forms of silence are ambiguous, leaving people to imagine the worst about the nonresponsive party.

FUNCTIONS OF NONVERBAL BEHAVIOR

Nonverbal behavior serves numerous functions, some of which have important implications for relationship building and creating trust in virtual work environments. For example, nonverbal cues may contradict what is said. If someone tells you that nothing is wrong in a sharp tone of voice, with arms crossed, and brows furrowed, the nonverbal cues contradict the words. Non-verbal cues are considered more believable than words when they contradict them.

Nonverbal cues are often looked upon to detect dishonesty in the sender of a message. Most people overestimate their ability to tell when someone is lying. Popular wisdom says that if someone is lying, it will leak out in nonverbal behavior. While this is true, most people don't look for the leaks in the right place. They believe that they can look in a person's face or eyes and tell if that person is lying, but typically a person who is being dishonest will focus on controlling his or her facial expression and eye behavior. (The people lying know that the face is what the people listening will probably be watching.) As a result, other body movements will betray the dishonesty because they are often overlooked. For example, people will use fewer illustrators when lying.[22] Some groups of people, most notably the U.S. Secret Service, are much better than average at detecting liars, indicating that, with some training, people can improve their skill in this area.[23]

The reliance on nonverbal cues to detect dishonesty may explain the importance of nonverbal behavior in the development of trust and the evaluation of credibility. Perhaps this, then,

---

Virtual Wisdom ▼

The other day, a salesperson told me she would do whatever it took to make me happy—and that really ticked me off! Why? Shouldn't I be thrilled at her offer? I wasn't because her words didn't match her body language.

—Anne Warfield, CEO of Impression Management[24] ▲

explains why a recurring theme in the management literature about virtual organizations is the challenge of building trust.

Nonverbal cues also serve an important function in the development of relationships. We can tell whether people like us by how much they pay attention to us. Do they look at us when we speak? Do they exhibit immediacy cues? Immediacy cues are nonverbal behaviors that communicate liking and interest. For example, if Megan is talking to Geoffrey about a new idea and Geoffrey maintains eye contact, smiles, nods his head, and leans a little toward Megan, then she will feel as if Geoffrey likes what she is saying and appreciates her. Megan would have a different feeling if he were looking around the room, watching other people as they walk by, not smiling or nodding, and leaning away from her. Not only would Megan surmise that Geoffrey was not interested in her ideas, she would probably also feel that he wasn't very interested in her as a human being either.

It is clear that some of the richness of everyday communication must be sacrificed for a virtual work environment that relies on asynchronous, written communication. Rich face-to-face interaction has no simple and efficient written counterpart. The trust and relationship-building roles of nonverbal communication may be the most difficult to replace in a virtual workplace. However, other forms of nonverbal behavior can be very useful substitutes for those that are missing.

- **Time.** How we use and respond to time is a powerful form of nonverbal communication. Meeting deadlines and being responsive to messages in a timely manner can demonstrate trustworthiness and liking.
- **Typography cues.** Typography such as boldface or italics can be used to show emphasis in writing as a substitute for paralanguage. Of course, typography can't capture all the subtleties of paralanguage, but it can be used effectively to emphasize a word or point.
- **Doing good work.** Off-site employees, managers, and virtual team members do much to communicate their trustworthiness and liking and make strides in relationship building simply by being dependable and performing well.

DIFFICULTIES WITH NONVERBAL COMMUNICATION

Limited face-to-face time or infrequent videoconferencing is no reason for remote workers or virtual team members to ignore the challenge of developing good nonverbal communication skills. Often unintentional, perhaps even subconscious, nonverbal behaviors have negative consequences. For example, odd eye behavior, such as shifting eyes, compromises credibility. High blink rates suggest stress. Poor posture communicates nervousness or lack of self-esteem. Nonverbal behavior that contradicts a verbal message creates a lack of trust in the message. While appropriate nonverbal behaviors can help showcase a manager's strengths, inappropriate behaviors can damage the perception employees have of a manager.[25]

Though nonverbal behaviors can add richness to verbal messages and help people understand them, people often misinterpret and misuse them. A simple gesture, such as crossing one's arms, can be interpreted as disinterest, when it may simply mean that the room is chilly. In an effort to make a good first impression, we may display high energy by offering a firm handshake and talking fast with numerous gestures. This attempt would likely backfire, however, because a slower, more controlled manner is associated with self-confidence.[26]

Virtual Wisdom ▼

Very often, management focuses only on such business needs as raising capital, procuring equipment, establishing operational plans, and providing technical training. Absorbed with those complexities, senior managers ignore or overlook the vital cultural issues that influence the day-to-day operations of their joint venture. To their chagrin, they eventually learn that the inability to manage a culturally diverse workforce usually affects every process in the chain, from development to marketing. Bureaucratic inefficiency, lower profitability, even insolvency, are the common results of the rising tide of frustration.

    —Dr. Farid Elashmawi, Global Success[27] ▲

The most important consideration of all is this: Interpreting nonverbal cues is really difficult. We often miss cues that are plainly there. We sometimes react to cues that no one else sees or hears. And, because of our age, our culture, our education, or a dozen other reasons, we often simply misinterpret what we perceive. We're not nearly as good at this as most people think, and the risk for successful communication is huge.

COMMUNICATING ACROSS CULTURES

As the world opens up to organizations, communication increasingly occurs across cultures. Virtual organizations and virtual teams may have to bridge cultural differences at many levels, from different functional areas within the same organization to different industries in different countries. Instilling a sense of cultural awareness is critical for team success, particularly with global virtual teams.

Virtual Wisdom ▼

Perhaps the most dangerous moment is when you think you understand. In the west, I sometimes think we focus too closely on words and their meaning, losing sight of the great range of nonverbal communications that also shape our pattern of contact.

    —Terence Fane-Saunders, crisis management expert[28] ▲

Though it may be the first thing that comes to mind, cultural differences include much more than just language. The subtleties that are often unspoken are just as important to understand. For example, individuals from different cultures vary in the degree to which they rely on the context of a message for cues to its interpretation. In high-context cultures, context has a considerable role in the interpretation of a message, while in low-context cultures the interpretation of messages depends more on the words themselves. Individuals from low-context cultures typically lack skill in noticing and interpreting contextual cues. These individuals may be perceived by those from high-context cultures as less knowledgeable and less trustworthy, because they frequently

Figure 2-2   **Context Scale of Different Cultures**

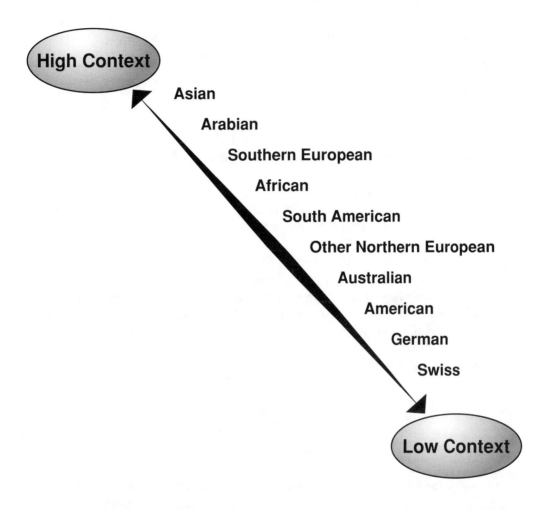

Source: Steven Beebe, Susan Beebe, and Mark Redmond, *Interpersonal Communication,* 2nd ed. (Boston: Allyn and Bacon, 1999), 109.

violate the high-context culture's unspoken rules for communication and behavior.[29] Figure 2-2 shows the world's major cultures on a scale from high to low context.

Robert Pabst, former president of Mark IV Audio, has extensive experience doing business with other countries. He illustrates the importance of context with his comments about doing business in Japan, a high-context culture known for its indirect style of communication. "As many Americans have learned," said Pabst, "it is difficult in Japan to know whether or not there is agreement. It seems that an obsession with politeness prevents definitive statements that may hurt someone's feelings. You may hear comments such as 'that's nice,' 'very interesting,' and 'I understand' during your presentation, leading you to believe, erroneously, that there is agreement.

Agreement is actually reached by talking, then listening, talking then listening, and so forth until both sides are saying the same thing."[30]

Pabst believes that U.S. business schools should commit to creating cultural awareness and educating Americans about cultures that they might encounter in the global economy. According to Pabst, "Most Americans grow up *knowing* that our way is the right way and that any other way is . . . well, simply wrong. Carrying this idea into business dealings in another culture can produce sad results, yet this is a common error." He offers the following suggestions for dealing with cultures from other parts of the globe.[31]

Be ready to contend with business methods you are prevented by law from practicing. For example, while the U.S. government condemns the two-martini lunch and penalizes businesspeople for entertaining customers, in countries like Japan, lavish entertainment is an expected part of an executive's compensation. Management salaries in Japan have been relatively low with the expectation that enjoyment of expensive restaurants, exclusive clubs, theater attendance, and other company-paid entertainment would offset the shortfall. In some countries, paying commissions, finder's fees, and the like to participants in a deal is essential to doing business. While this is a normal and accepted practice in these societies, paying such a fee can land you a stay in a U.S. federal prison. Know how the law conflicts with business practices you face in another culture and be prepared with a means to overcome the conflict.

Informality has enhanced the success of U.S. domestic business but in some other cultures, it can doom an effort to failure. The clothing you wear, the seating arrangement around a conference table, and other considerations of this sort may not even be noticed in a meeting of Americans, but they may have a strong influence on the outcome of a meeting with people from other countries. In rural Germany, for example, even a husband and wife are likely to address and refer to each other as Herr Schmidt or Frau Schmidt in a business-related setting rather than use first names. Certainly, business associates are expected to avoid using first names until after an extended acquaintance and then only with permission.

Some formalities can seem a bit bizarre. Pabst once visited a business prospect in Romania, rising bright-eyed and alert for an early morning meeting. As attendees gathered and took the appropriate seats around a long conference table, he noted that two bottles of liquor had been placed on the table's center: Cutty Sark scotch and something that looked like vodka with a label he couldn't read. Pabst soon learned that it was a formality there to open a business meeting with a drink—at 8:00 in the morning! To be polite, he chose the clear liquid, wanting to show an interest in the local culture. After downing the shot, he was unable to talk for several minutes, concluding in retrospect that it was some form of petroleum distillate. He did pass the formality test, though, and the meeting ended successfully.

Be aware that what they say isn't always what they mean. In an Asian country, Pabst had been confused by several breaches of what he had thought were agreements. His tolerance ran out when, after negotiating an exclusive manufacturing agreement, he discovered that a nearly identical product was for sale at a trade show in Europe. Pabst challenged the manufacturer and the product was withdrawn from the market, but he had the impression that their agreement was somehow not taken seriously until he challenged the trade show discovery.

A good rule here, but an essential one over there, is to always give the other person a graceful way out. You may have heard that people from Asia are concerned with *saving face*. Note well that this is very serious, and not just with Asians. Recall the emotion that may grab you when you think you know the answer in class but can't raise your hand for fear of being wrong, being embarrassed, and *losing face*.

A group of German salespeople had proposed a particular distribution scheme for Pabst's approval. He knew from experience that the proposed method would not perform as he had hoped, despite its attractiveness on the surface. Pabst was explaining the flaws in their proposal and was about to be quite critical when he noticed that a distinct chill had settled over the meeting. He called a break, asked an associate what was going wrong, and learned a lesson in saving face—he had to give them a way to not be embarrassed or feel defeated as the proposal was turned down. When they reconvened, he was able to lead the discussion by asking some what-if questions that allowed the group to see for itself that they shouldn't go ahead with the proposed scheme.

The American approach of "cut the chatter and let's get down to business" doesn't fly in some cultures. In virtually all nations outside of North America, there is an extensive warm-up period before serious discussions begin. This may be at the meeting site and include get-acquainted discussions, plant tours, and presentations about the company's products and markets; or it may be a city tour, dinner or lunch, some form of entertainment, and then the routine at the business site. You cannot hurry this process, and you shouldn't try.

This latter point lends support to a general rule that Robert Pabst urges Americans to observe when conducting business with people from another culture. It is to take things slowly while watching and listening to detect what behavior is expected from you.

Global virtual teams are especially likely to use e-mail to work across space and time. This can be an advantage when communicating across a language barrier. Very often people feel they can write in their nonnative language better than they speak it. Still, when composing written messages for receivers from other countries, take extra care to ensure effective communication.[32]

- Be concise and precise.
- Use specific words and concrete examples to explain your points.
- Avoid slang, jargon, and buzzwords. Phrases like "off the wall" or "tuned out" can be really confusing to someone from another culture.
- Use short sentences and paragraphs.
- Raise the level of formality slightly.

General suggestions are only so helpful. Communication with co-workers and teammates in other countries will be most improved by researching the communication norms for their specific cultures.

DISCUSSION QUESTIONS

1. What are some examples of noise that are frequently experienced in the workplace? What suggestions do you have for overcoming these forms of noise?

2. How would you feel about working with someone that you had never seen? How would this affect your communication? Is there a benefit to posting photos online of distributed team members that will not meet? Is there a reason why you may not want to do this?

3. Psychologist and philosopher William James believed that there are three components to the self: the social self, the spiritual self, and the material self. The social self refers to who you

are when you are interacting with others. The spiritual self is composed of your thoughts and feelings about your values and morality. The material self is the total of all of your possessions, including your body. If someone were to enter your room, what conclusions would he or she draw about you based on your material self?

4. What do you believe would be the most difficult aspect of communicating with co-workers primarily without nonverbal cues? Is there a limit to how much relationships can develop without nonverbal communication? If so, how do you explain romantic relationships that develop online?

ENDNOTES

1. A. B. Rami Shani and James B. Lau, *Behavior in Organizations* (Boston: McGraw-Hill Higher Education, 2000), 303.
2. Steven Beebe, Susan Beebe, and Mark Redmond, *Interpersonal Communication,* 2nd ed. (Boston: Allyn and Bacon, 1999), 11–16.
3. James S. O'Rourke, IV, *Management Communication: A Case-Analysis Approach* (Upper Saddle River, NJ: Prentice Hall, 2001), 21–22.
4. William R. Pape, "Hire Power," *Inc.,* 2002. Available: *http://www.inc.com.*
5. Manju K. Ahuja and Kathleen M. Carley, "Network Structure in Virtual Organizations," *Journal of Computer-Mediated Communication,* vol. 3 (1998). Available: *http://www.ascusc.org/jcmc/vol3/issue4/ahuja.html.*
6. Laurie Putnam, "Distance Teamwork," *Online,* vol. 25 (2001): 54–58.
7. Wayne Cascio, "Managing a Virtual Workplace," *Academy of Management Executive,* vol. 13, no. 3 (2000): 81–90.
8. John V. Thill and Courtland L. Bovee, *Excellence in Business Communication* (Upper Saddle River, NJ: Prentice Hall, 1999), 26–30.
9. Cascio.
10. Martha Haywood, *Managing Virtual Teams: Practical Techniques for High-Technology Project Managers* (Boston: Artech House, 1998).
11. A. Mehrabian, "Verbal and Nonverbal Interaction of Strangers in a Waiting Situation," *Journal of Experimental Research in Personality,* vol. 5 (1971): 127–138.
12. Bavelas and Chovil.
13. James P. T. Fatt, "It's Not What You Say, It's How You Say It," *Communication World,* vol. 16 (June/July 1999): 37–41.
14. Paul Ekman, "Facial Expression and Emotion," *American Psychologist,* vol. 48 (1993): 384–392.
15. Peter Guiliano, "Seven Benefits of Eye Contact," *Successful Meetings,* vol. 48 (1999): 104.
16. Fatt.
17. L. E. Tyler, *The Psychology of Human Differences* (New York: Appleton, 1956).
18. A. Feingold, "Good-looking People Are Not What We Think," *Psychological Bulletin,* vol. 111 (1992): 304–341.
19. Anne Warfield, "Do You Speak Body Language?" *Training and Development,* vol. 55 (2001): 60–62.
20. Lisa Kimball and Amy Eunice, "The Virtual Team: Strategies to Optimize Performance," *Health Forum Journal,* vol. 42 (1999): 58–63.
21. John Perry, "Palm Power in the Workplace," *American Salesman,* vol. 46 (2001): 22, 5p.
22. Paul Ekman and Maureen O'Sullivan, "Who Can Catch a Liar?" *American Psychologist,* vol. 46 (1991): 913–920.
23. Dave Zielinski, "Odd Language Myths: What You Think You Know About Body Language May Be Hurting Your Career," *Presentations,* vol. 15 (2001): 36, 7p.

24. Warfield.
25. Ibid.
26. Zielinski.
27. Farid Elashmawi, "Overcoming Multicultural Clashes in Global Joint Ventures." Available: *http://www.iftdo.org/articles.htm.*
28. Terence Fane-Saunders, "Public Relations in Asia: The International Dimension," 22 February 2002. Available: *http://www.issa.com/features/intl/2002feb22.html.*
29. Beebe, Beebe, and Redmond.
30. Robert Pabst, personal communication with author, February 2002.
31. Ibid.
32. Thill and Bovee.

3 TECHNOLOGIES FOR VIRTUAL ORGANIZATIONS

"Ideas," according to Alison Overholt of *Fast Company* magazine, "need to move faster than ever. Global teams have to cooperate more closely than ever. Nonstop travel seems less appealing than ever."[1] She offers a solution in the form of tools for electronic collaboration. Virtual organizations, virtual teams, and telecommuting are made possible by the development of advanced communication technologies. And, in this chapter, we'll discuss some of the options available for electronic communication and suggest ways to use them most effectively. Keep in mind that advances in communication technologies outpace the printing of textbooks, so what you will find here is a general overview of categories of technologies. If you want to learn about the newest applications or about specific products, a technical periodical would be a better resource than a book. Staying on top of the latest developments in communication technology requires a commitment to continually updating your knowledge.

As you investigate the most recent offerings in communication technology, be aware that new technologies will continue to develop, and you should consider issues of speed, reliability, availability, cost, and ease of use in addition to their features. Cost is a particularly important consideration for most managers, and communication technologies can represent a significant expenditure, even though the motivation to create virtual work environments often is to save money. Dr. Laura Esserman of the University of California–San Francisco's Carol Franc Buck Breast Care Center cautions purchasers of virtual technologies to avoid being limited by a "cost-cutting mindset" that can eclipse the true potential of these tools to enable people to "work smarter" rather than just cheaper. She uses communication technology to connect patients to the best information and advice from a roomfull of doctors, instead of just one. This wealth of expertise helps patients who are given a difficult diagnosis (such as breast cancer) that requires them to make a number of tough choices (e.g., lumpectomy or mastectomy) to choose what is best for them.[2]

Dr. Esserman recognizes that doctors frequently recommend treatment programs that they are most familiar with, but that those may not be the best ones for the patient in terms of long-term physical and emotional consequences. Uninformed patients are likely to follow the doctor's recommendations because they don't know enough about alternatives to feel confident in their own choices. Dr. Esserman is working to change this with the help of "decision community" software developed by a Pittsburgh company called MAYAViz. Through the use of this software, a patient is given a printout of treatment options with risks and benefits, statistics, and physicians' experiences with each option. Patients can ask questions about treatment options, and, if their physicians have limited experience with a given treatment, they can electronically call on the wisdom of another physician. This allows patients to become fully informed about their options and to feel confident about their choices.[3]

Advanced communication technologies offer organizations numerous advantages, from saving money and time to reducing paperwork and anxiety over whether or not a document has been received.[4] In the following pages, we will explore various communication technologies, some newer and some tried and true.

BASIC TOOLS

E-MAIL

In many traditional organizations, the most common form of communication is the face-to-face conversation. In virtual work environments, the most relied upon means of communication is e-mail. One major difference between these is synchronicity. Communication technologies used in the virtual workplace can be synchronous (occurring at the same time) or asynchronous (not occurring at the same time). Each type has its advantages and disadvantages. Synchronous media, such as phones or electronic chat, are convenient when the sender of a message wants to check that the receiver has understood the message, but they require the sender and receiver to be available at the same time. For virtual teams, particularly global virtual teams that work across time zones, this mutual availability can be difficult to manage. Asynchronous media, such as e-mail and electronic bulletin boards, allow senders to send messages at their convenience, but they have little control over when the message is actually received.

E-mail is by far the most popular form of computer-mediated communication. It is used in offices, on campuses, and in homes all over the world for sending professional and personal messages. In the workplace, e-mail is an exceedingly useful tool that has some tremendous advantages over other methods of communication. First and foremost, it's quick and easy. It's asynchronous, so you may use it at your convenience to send a message. You can reach multiple audiences by sending only one message. You can transfer electronic files to others by attaching them to an e-mail message. And you can keep a record of messages that you have sent and received.

In many ways, e-mail has changed the way we communicate even in a traditional office. It offers an informal alternative to face-to-face or phone communication. Before e-mail, if you had to deliver an informal message to someone in an office down the hall, you had two choices: You could physically walk down the hall and deliver the message face to face (if the person was in his or her office), or you could call. Each of these synchronous communication options has some possible disadvantages.

- **Synchronous communication can be inefficient.** Most of us have had the experience of wanting to give a brief message to someone who really likes to talk. Brief messages can turn into lengthy conversations that can eat away at the day, sometimes without our awareness. With e-mail, the sender of the message is in control of how much time is allotted to completing the message.

- **Synchronous communication can be disruptive.** If you are in the middle of a demanding task, you may not want to have your work disrupted by someone at your door or by a ringing telephone. In addition to the disruption itself, getting refocused afterward can take some time. When you are on the receiving end of an e-mail message, you have control over when you accept the message, so your work doesn't have to be interrupted.
- **Asynchronous communication is preferred by some personalities.** People with more introverted personalities may prefer to communicate by e-mail than face to face or by phone. These people are more likely to communicate when e-mail is available.

E-mail is the lifeline for virtual organizations and virtual teams. It is a cost-efficient way to communicate across the miles and an effective way to communicate across time zones, without losing sleep. E-mail is the most common form of communication for remote employees. It is a unique medium in that it is written yet is still quick and can feel very personal and conversational. These qualities make e-mail an integral part of the virtual workers' technical connection to their job and their human connection to the organization and other workers.

Of course, e-mail does have its disadvantages. With e-mail, most nonverbal cues are lost, which can lead to more frequent misunderstandings. With virtual workers, the context of a message also may be unclear. For example, if there is a great deal of excitement around an office over a certain event, a remote worker who is not aware of this fervor may be confused about the meaning of a message created in this context. In addition to these disadvantages, your e-mail message looks different to the receiver than it does to you when you send it. Charts, tables, bulleted lists, and some typography don't always transmit well. Some downloads fail to convert.

It is also important to realize that e-mail is not private. Using company-owned hardware and software for e-mail means that the company has every right to monitor an employee's e-mail, unless an expectation of privacy for employee e-mail has been expressly created.

For telecommuters working at home, it may seem that the issue of e-mail privacy is not as great a concern as for those employees in a call center or traditional office. But e-mails sent by remote workers to their co-workers and team members in traditional offices may well be seen by others in the organization. In one case, the sales manager at a small, privately owned Midwestern company was sending the controller, who was a friend, casual e-mails that contained jokes along with some disparaging remarks about the owner of the company. The controller read the e-mails and deleted them, but failed to empty the trash folder. One day when the controller was home on a personal day, he received a call from the owner asking for his password so the owner could access some financial data. The owner never said that he intended to read any of the controller's stored or trashed e-mail messages, but the following day when the controller returned to work, the sales manager was fired for sending the disparaging e-mails.

Monitoring e-mail can be a matter of routine, or it can be the result of a direct investigation of suspected inappropriate e-mail. Many organizations have begun to monitor e-mail over concerns about possible lawsuits in which e-mail messages may be used as evidence. In a 2001 survey of 435 firms, 10% reported having employee email subpoenaed.[5] As surveillance software and other monitoring tools become less expensive, more companies are beginning the practice. In 1997, only 14.9% of companies reported storing and reviewing employee e-mail messages. By 2001, that figure had jumped to 46.5%.[6] Very often employees make the mistake of assuming that e-mail is private, only to be shocked and dismayed when they discover that their e-mail has been read by others in the organization. This is, in part, because many companies have no formal e-mail policies and those that do often don't do a good job of making them known.

Once e-mail messages are sent, the sender has no control over where they are forwarded or where they are stored. Neal Patterson, CEO of Cerner Corporation, learned this when he sent a rather scathing e-mail to around 400 company managers. He had been disappointed by the nearly empty parking lot outside his Kansas City office at 7:30 a.m. and 6:00 p.m., which he took to reflect a lax work ethic. He blamed his managers for creating "a very unhealthy work environment" and promised to hold them accountable. In his e-mail, he listed punishments he planned to implement for the Kansas City employees, including having the employees punch a time clock and possibly laying off some people. He wanted to see the parking lot "substantially full at 7:30 a.m. and 6:00 p.m. and on Saturday morning." He gave his managers two weeks to turn the situation around and ended his e-mail with an ominous "Tick, tock." After clicking "send," the e-mail left Patterson's control and took on a life of its own that ended with it being posted on Yahoo!, where anyone could read it, including analysts and investors. Readers took the harsh message as an indication that something was very wrong at Cerner. In the days after the leak, the trading volume for Cerner surged and the stock price plummeted.[7]

In some cases, e-mails are shared with co-workers purely by chance. A new employee at a small company had been using the company's e-mail to send romantic messages. What she didn't know was that when the e-mail system went down, the last e-mail message sent by anyone in the system before the crash would be automatically printed when the program came back up. She learned this the hard way when the e-mail message that printed automatically when the system came back up after a crash was a rather suggestive note she had written to a man other than her husband.

USING E-MAIL STRATEGICALLY IN THE VIRTUAL WORKPLACE

Composing e-mails deserves attention in this chapter because e-mail is the primary means of communication for remote workers. In a virtual organization or virtual team, while it is preferable for team members to meet each other face to face when the team is formed, this is not always feasible. It is possible, then, that those communicating by e-mail may have no idea what the person with whom they are communicating looks like or dresses like, or what the person's office is like, or how that person is perceived by others in the organization. In multi-organizational teams, team members may know each other's job titles, but that may not clearly communicate much about the person's position within his or her organization.

In these circumstances, much of the information that we traditionally use to form impressions of others is missing. When communicating by e-mail, many nonverbal cues are absent. Thus, people are forced to form impressions using the few cues left available to them, such as the timeliness of the response to messages. For example, a prompt response to a message communicates interest and professionalism.

The content holds valuable information for impression formation, both in the meaning of the message and in how the message is prepared. For example, a message that contains a good solution to a problem communicates competence. However, if that message also contains several typographical errors, the message may communicate carelessness. With online communication, your professionalism, your competence, your likability, and your trustworthiness are communicated through your writing.[8]

Many people have taken the speed and efficiency of e-mail to heart but have neglected to recognize that the messages can communicate beyond the content of the words. Messages that are unclear or that contain spelling or grammatical errors seem unprofessional. Still, there is much to be said for not compromising the cost effectiveness of e-mail by spending too much time on each message. The amount of time spent proofreading any e-mail message should be brief, in part because the message itself ordinarily should be brief. The time spent revising your e-mail messages should depend on your communication objective. If a message were very important, then your objective would be for the message to be taken seriously. If the receiver of the message is a professional contact whom you've not actually met, your objective might be to create a good impression. If the message were for an external audience, part of your objective would be to represent your organization positively. In these and similar situations, a more meticulous review of your message would be warranted. In writing e-mail, seek a balance between spending so much time on an e-mail that you compromise its cost effectiveness and being hasty with e-mail and compromising your career.

Virtual Wisdom ▼

The written word can be so much more harsh than the spoken word; even a critique needs to be phrased positively.
—Karen Davidson, President
KDA Software[9] ▲

When composing and revising e-mail, think about how your message will read to your audience without your body movements, facial expression, or tone of voice to add clarity. It may be difficult for the reader to know how to take something you've written without the normally accompanying nonverbal cues to help with interpretation. For this reason, it is advisable to be very cautious when using humor with people who don't know you, as it could be misconstrued as offensive. Humor is a useful tool, just be careful.

You can use typography cues to create emotion that normally would be communicated nonverbally. Consider the following message:

I need your response by Thursday.

The meaning of this message can be altered considerably without changing a single word.

I need your response by THURSDAY!

I NEED YOUR RESPONSE BY THURSDAY!

I need *your* response by Thursday.

I **need** your response by Thursday.

I need your response by Thursday :-)

Typography cues, such as bold and italics, can be used to create emphasis and direct readers' attention, but they should be used sparingly. Too much variation in typography can be distracting. And in some cases, the receiver of your message may not be able to see it correctly with his or her e-mail system. To be sure that you're sending typography cues that your receiver will be able to read, you can use punctuation and capital letters. But use all caps in extreme moderation. Typing in all caps is the equivalent of shouting and is considered flaming (abusive behavior) and is not appreciated by the receiver of a message. In addition, reading all caps is simply difficult for the eye. All caps should be used only for the occasional, individual word that you want to emphasize.

Emoticons are typographical symbols of emotions. In some virtual communities emoticons are quite popular, and in others they are rarely seen. Emoticons may be considered unprofessional, so you should avoid using them in workplace communication unless you have a personal as well as a professional relationship with your receiver.

Many people use abbreviations that have emerged from e-mail and chat rooms to lessen keystroking. However, these too may be seen as unprofessional and can frustrate people who do not know what they mean.

Popular Emoticons		*Less Common Emoticons*	
:-)	a smiley	{:-)	a smiley with a toupee
;-)	a winking smiley	::-)	a smiley with glasses
:-(	a frowny	:-#	a smiley with braces
:-o	a surprise	:-`	a smiley spitting out chewing tobacco
:->	a sarcastic smiley	(8-o	a Mr. Bill smiley
		~(_8^(l)	a Homer Simpson smiley

Common Abbreviations			
ADN	any day now	OTW	on the whole
B4N	bye for now	SOP	standard operating procedure
FWIW	for what it's worth	WB	welcome back
IAE	in any event	LOL	laughing out loud

TIPS FOR USING E-MAIL EFFECTIVELY

- **Be concise and direct.** Busy professionals may be frustrated by long e-mails that take numerous lines to get to the point, and they may not read the entire e-mail. In her book, *Guide to Electronic Communication,* Kristen Bell De Tienne recommends keeping e-mails to one screen if possible. If the message must fill more than one screen, she recommends previewing all the main ideas on the first screen.[10]

- **Use a specific subject line.** Michael Dell receives around 200 business and personal e-mails a day. "I'm constantly checking my e-mail, and getting behind on it can be overwhelming," said Dell.[11] When people receive a lot of e-mails, especially busy people, they may scan the subject line to decide which e-mails to read right away and which e-mails to read later or to delete. To help reduce the volume of incoming e-mail, add "No reply requested" at the end of messages. (Or, conversely, state that a reply is requested to let your reader know that you definitely want feedback.)

- **Use a signature block.** For external communication, create a signature block that is automatically placed at the end of your e-mail messages that includes your name, title, address, phone, fax, and web site address.

- **If you know you want to keep a message, print it.** More than one person has lost a year or more of back e-mails due to unforeseen technical problems. Online or Zip Drive storage may be suitable alternatives to paper, but for them to be useful, you must back up regularly. This is especially important for remote workers who have only an electronic inbox, rather than one full of hardcopies that could be stored.

- **Delete original messages before replying to them.** If the original message or accumulation of messages is long, it can take longer to download. Do not forward messages that have been forwarded multiple times without deleting the header information from all the previous forwards. The receiver of the message may get annoyed by having to scroll though all of this to get to the message.

- **Do not forward chain letters.** Please. We beg of you, don't do it.
- **Scan attachments for viruses before opening them.** Your computer can become infected with viruses from attachments to e-mails. The virus infects your computer when the attachments are opened.
- **Never compose and send e-mail when you are angry.** "I make sure anything in an e-mail is neutral," says David Chalk, Chairman of Chalk.com Network, Inc. "If I get an angry letter from someone about anything, I don't reply by e-mail ever, ever, ever."[12] Once you click the send button, the message is gone. You can't recover it later when you've calmed down.

For distributed teams, it may be useful to create a team address or a mailing list for the entire team and to encourage team members to use the mailing list frequently when communicating. This ensures that information is effectively shared across members.[13] Be aware, though, that with some e-mail systems, when you reply to a mailing list, your reply will go to everyone on the list, not just the sender of the message. When sending a message to a mailing list, the individual e-mail addresses of those on the list are not shown. This eliminates any concern about the order of the appearance of the addresses. It may seem like address order should not be a concern, but some people have read meaning into the position of their address on a recipient list. If you must use a list of addresses instead of a team address or if you want to show the addresses, arranging the list of recipients alphabetically or in order of seniority is a way to avoid misperceptions about the significance of the order. Keep your original message from any mailing list you create or join. It will give you the procedure for being removed from the list, which may be important information for you at some point in the future.

TELEPHONES AND VOICE MAIL

The explosion in the use of e-mail hasn't eliminated the need for phones. Even though e-mail is the most relied upon means of communication in a virtual workplace, there are times when the phone is more appropriate. Rodney Heeden, President and CEO of Relizon Company, a provider of software and management services, thinks that e-mail is sometimes inappropriate because it seems impersonal. "If I have a message I want to send to every employee, I use voice mail because I think it's important that they hear my voice and my tone," he explained.[14]

Today professionals have a plethora of choices when it comes to phones, and today's phones can be used to do much more than make calls. Smart phones combine the capabilities of cell phones and PDAs in one device. According to Dave Johnson of *Laptop* magazine, this is a sensible development given the rise in cell phone and PDA use. Just maintaining multiple gadgets and keeping all their batteries charged can be a challenge. Smart phones offer users the convenience of carrying only one device, and much more. With a smart phone you can dial numbers from the contact list on your PDA, send and retrieve e-mail, access the web, and log into the company intranet, all from the road. These devices can keep the telecommuter who leaves home to see clients or go to a parent-teacher conference connected to the office.[15]

TIPS FOR USING THE TELEPHONE EFFECTIVELY

- **Avoid multitasking while on the phone.** It is often tempting to do other things while speaking on the telephone. You should avoid this temptation when you are engaged in professional communication, even when you are at home. If you work from home, this may not be possible all the time; someone may call while you are in the middle of an activity that simply cannot be stopped in its tracks. In that case, you should ask to call the person back, rather than attempt to multitask. Even though you may think that you can pay careful attention to a phone conversation while you do other things, the person on the other end of the line may be aware that you are doing something else (e.g., they hear the clicking of the keys as you input data). This can make for an unsatisfying interaction, leaving the person feeling as if he or she did not warrant your undivided attention.

- **Make certain the phone is answered professionally.** Remember that if you work from home and you do not have a separate line for business, your personal answering machine message represents you to the clients and colleagues who call when you are unavailable. An unprofessional message can affect the confidence that others have in you, especially when they don't know you well. Coaching your children on their telephone skills may also be a good idea if they occasionally answer business calls.

BULLETIN BOARDS AND WEB PAGES

Electronic bulletin boards and web pages offer places for users to share information by posting messages, documents, or graphics. These spaces can be accessible to all members of a virtual organization or team and can be made available to anyone else the team chooses.

A bulletin board can be a useful tool for gathering information and ideas from a large, diverse group of people who are not necessarily part of the team. For example, one Midwestern college with a number of satellite campuses creates continuity in the coursework at the various campuses by having a single textbook and a master syllabus for all their core courses. The master syllabus serves as a guide for all faculty who teach the course. One faculty member is asked to create a draft version of the master syllabus. That version is then posted on the web, and a bulletin board is created where any faculty member from any campus can make comments and suggestions after viewing the draft. Visitors to bulletin boards can view a record of the information previously posted, follow the thread of the postings, and respond to and build on the ideas of others.[17]

Web pages are an effective and efficient way to share information with virtual team members and to those outside the team. However, some web pages are better than others. The most important consideration with a web page is its design.

TIPS FOR CREATING EFFECTIVE WEB PAGES

- **Make the site easy to navigate.** An effective web site is well organized and easy to navigate and provides all the needed links to other parts of a site. Design elements such as color, contrast, repetition, alignment, frames, and proximity should be used to create the organization of the material and guide the reader's eye. When planning the site, consider the audience and how they will use the site, and then organize information in a way that will make sense to them. Few things are more frustrating than trying to get information from a poorly designed site, such as one that doesn't offer important links (internal and external) or one that doesn't include enough headings to break up text and make things easy to find. The home page, or the first page that you see when you enter the site, should direct visitors to wherever they want to go.[18]

- **Make the site visually appealing.** The appearance of the site should be appealing and communicate something about the content of the site. For example, an organization can use font, color choices, and graphics to communicate that the content of its site is serious and the organization is conservative. Visit *benjerry.com,* the web site for Ben & Jerry's, to see an example of a site that communicates organizational culture visually.

- **Use design elements strategically.** Paragraphs should be brief and relevant graphics should be included. Avoid using design elements simply because you can. You will wind up with a page that is too busy, and all the bells and whistles can be distracting. Adding multimedia (graphic, audio, and video files) to your site can add some value, but they do slow things down a bit.[19] If your site will be accessed by off-site workers, keep in mind that telecommuters may not have huge bandwidth available in their homes, and a page with a lot of multimedia will take a very long time to load.

REMOTE ACCESS SOLUTIONS

Remote access software and services are wonderful tools for individuals who telecommute only part of the time and maintain a computer at the office as well. Until recently, if you wanted to access your files at the office, you could either physically go to the office or carry a laptop to and from the office. Remote access systems allow you to work on the computer in your office from a remote location. These systems allow you to sit at your remote computer (the guest computer) and control your office computer (the host computer). This means that your office files are updated and synchronized as you work. These functions can be implemented through either software-based systems (better for the telecommuter who uses only one guest location) or web-based systems (better for the telecommuter that visits clients and may use numerous remote locations).[20]

ALL-IN-ONE PRINTERS

All-in-one printers that print, scan, copy, and fax are great basic tools for the telecommuter. Although you can attach a document to an e-mail, there are times when an electronic copy of a document will not suffice. Copies of signed documents, documents with handwritten markings or editing, and documents to which you have no electronic access may be rapidly transmitted with the use of a fax machine.

On the other hand, you may want to scan the document to create an electronic copy. Or you may want to make numerous hardcopies to send to various locations. Regardless of the specific situation, it is much more convenient to have the capability to scan, copy, and fax from your home office than to have to run to a full-service copy shop.

TOOLS FOR MEETINGS

Virtual meetings are becoming more commonplace as companies cut down on the amount of time and money allotted to travel. Since the September 11 terrorist attacks, safety has also been an issue and has fueled an interest in electronic meeting technology. This industry grew from 30% before the attacks to 40% in the months after.[21] This new technology, which until recently was often rejected in favor of hopping on a plane, is now gaining acceptance.

Virtual Wisdom ▼

If you have the right training, then technology will be your friend.
—Robert Reich, former U.S. Labor
        Secretary[22] ▲

AUDIOCONFERENCING

Audioconferencing, or conference calling, is a familiar tool and useful for virtual teams. It provides a synchronous communication environment, where immediate feedback and perception checking are possible, and it allows for the transmission of some non-verbal cues, such as vocal tone. Elizabeth Allen, Vice President of Corporate Communication at Dell Computer, communicates with her virtual team members primarily by e-mail, but uses audioconferencing when they need to meet.[23] Audioconferencing does have a few disadvantages, which can include cost. Another is the confusion that may result from trying to keep track of who is speaking. This is especially true with a larger group of participants.

TIPS FOR EFFECTIVE AUDIOCONFERENCING

- **Share your meeting agenda in advance.** As with all meetings, an audioconference will be more successful if you are prepared, share your agenda in advance, and set up and adjust all equipment early so you can start the meeting on time without technical difficulties. Allen says that at Dell she will "send out materials ahead of time" including a list of participants. If people see an unfamiliar name, they can look that person up on the Dell directory.[24] Distributing an agenda in advance is a good suggestion for all meeting formats.
- **At the beginning of the call, introduce everyone who is participating.** Receiving a list of participants in advance lets you know whom to expect. Introductions at the beginning let you know who showed up. If there are a number of participants, and they are not well known to each other, it can be useful to continue to introduce oneself each time one speaks. With a distributed group, time differences may be an issue and may warrant some acknowledgment at the beginning of the call.
- **Keep noises to a minimum.** If you are joining a teleconference from home, do all that you can to eliminate distracting and disruptive background noises, such as a barking dog, noisy children, or a television. The same concern exists for those on distributed teams calling from a noisy office. Ringing telephones and chatting co-workers can be picked up easily by a speakerphone and be distracting. A headset is less sensitive to this type of background noise.

- **Facilitate participation.** The larger the group participating in a conference call, the more difficult it is to ensure that everyone contributes. Some participants may be reluctant to speak and may need to be asked to share their views.
- **Use good equipment.** According to Elizabeth Allen, "This is very important." Just putting a group of people around a speakerphone isn't good enough.[25] A number of options are available for professional teleconferencing. With reservationless teleconferencing, a phone number is dedicated to the conference. Employees can dial that number and join a teleconference, making it possible to meet with very little advanced planning. Operator-assisted teleconferencing is a bit more formal. An operator greets participants and introduces them to other participants when they join the conference.[26]

VIDEOCONFERENCING

Videoconferencing can be a useful way for virtual teams and organizations to meet, although the quality can be disappointing. For people working from home and using a modem connected to a regular phone line with smaller bandwidth, images are not crisp and movements are jerky. For employees telecommuting from home, a desktop system would be used for videoconferencing. These systems use a desktop computer with videoconferencing software and a web camera. Video images are transmitted over regular phone lines and the web.[27]

Room systems for videoconferencing are more expensive because they require special cameras, hardware, and high-speed Integrated Services Digital Network (ISDN) lines. Because of the expense, these systems are more likely to be found on-site at an organization. The University of Notre Dame partners with the accounting firm of Ernst & Young, LLP, to offer their employees a Master of Science in Accountancy degree using room systems located at the University and various E&Y offices. Students in the various locations sit in a room together and view the professor, who delivers a lecture in front of a camera on the Notre Dame campus. By pressing a button on a plate or by simply speaking with a voice-activated system, students can ask questions; the camera will focus on them and the microphones will pick up their voices, so that they are both seen and heard by the professor and the students at other locations. A document camera is available to project papers and three-dimensional images to all locations, as well. Because of the high-speed lines, the quality of the images and movement is nearly as good as commercial television broadcasts.

Virtual Wisdom ▼

Videoconferencing was around long before e-mail, but it had all the appeal of a Beta-max player. Now, recent world events have brought it back into vogue.
—Sarah Bean, business writer[28] ▲

TIPS FOR EFFECTIVE VIDEOCONFERENCING

- **Consider your business objectives when making decisions about equipment.** Videoconferencing equipment is expensive and takes training to learn how to use. Alternatives include renting the equipment or leasing a videoconferencing suite. Another alternative is videophoning. With a computer, a microphone, a video camera, and the software from Eyeball Networks, up to six callers can appear on each other's screens simultaneously for a meeting.[29]

- **Know how to use the equipment and test it before meetings.** Because videoconferencing can feel clumsy to start, be sure to test all the equipment and make all adjustments before the scheduled time for the meeting.
- **Train the participants.** People are often uncomfortable in front of a camera and may not like having what they say recorded on video. Training can help ease their discomfort by increasing their confidence and managing their expectations.
- **Look at the camera.** Eye contact is important in face-to-face communication. Although people aren't actually face to face with videoconferencing, looking at the camera creates the illusion of eye contact. Looking at the screen image of the other participants in the meeting gives the impression of looking away from them. Remember also that you will be seen when the video camera focuses on you, so appropriate attire is required. Position yourself so that your image fills the frame, but not too close to the screen. Distance from the camera can communicate greater levels of formality and disinterest.[30]

> **Virtual Wisdom ▼**
>
> *It's the middle managers who may resist videoconferencing as it prevents them from getting more air miles.*
> —Martin Hill, Managing Director
> MVC[31] ▲

WEB CONFERENCING

Web conferencing combines audioconferencing with document sharing. This can be a very cost-effective tool for electronic meetings and collaborations. The most popular form at the moment is a presentation-centered version that combines audio with the ability to show PowerPoint slides. Conference service providers, or conference hosting services, offer organizations an inexpensive way to experiment with web conferencing. Organizations such as Centra Conference and WebEx will host web-based discussions for a fee. Those fees are based on time (mShow charges 50 cents per minute), or per seat (PlaceWare Conference Center charges $400 per seat), or some combination of the two (SameTime charges $5,000 per conference and $20 per seat).[32]

Organizations have found some surprising advantages to web conferencing above and beyond saving money and time. For example, Texas Instruments, which conducts meetings with participants all over the globe using WebEx, has found that its international attendees are more likely to offer feedback during web-based conferencing than when face to face.[33]

TIPS FOR EFFECTIVE WEB CONFERENCING

- **Do some research before choosing a product.** Choosing the best application for virtual conferences requires some investigation because different vendors offer unique features. *Fast Company*'s Alison Overholt points out the distinguishing features of four virtual collaboration tools. WebEx allows participants other than the host of the conference to take control of the desktop and change a document. Centra offers a flexible service that allows participants to link to other views, so, for example, a user can break out of a presentation to meet with a small group and then rejoin the presentation later. PlaceWare's Question Manager feature allows users to ask questions of a designated panel of experts during a presentation, offering two distinct advantages. Users can ask questions without everyone else in the group knowing, and they can do so without disrupting the presentation for everyone else.[34]

- **Stay focused.** Sitting alone in front of your desktop during a web conference, it is easy to mentally drift. To help yourself stay tuned in, avoid doing other things at your desk, such as sorting through mail or straightening a desk drawer. Stay active in the meeting by taking notes or asking questions.

GROUPWARE

Groupware is a category of software designed to meet a variety of needs for people working in collaboration. Groupware products were originally designed for asynchronous collaboration, so that groups could work together but not necessarily at the same time. Products like ERoom Digital Workplace offer important features for group work, such as team calendars, customizable databases, drag-and-drop file sharing, document version control, polling, and approval routing. Many of the groupware products have enhanced their offerings by adding synchronous features. For example, ERoom has an add-on product, called ERoom Real Time Server, with a chat feature so groups can communicate synchronously.[35]

Virtual Wisdom ▼

A product that works great inside a company can fail miserably when you are trying to connect through your firewall with someone on the outside.

—Mark Levitt, Research Director
Collaborative Computing[36] ▲

According to David Holan of Frogscape Digital Design and Consulting, the features available among the current groupware offerings vary from vendor to vendor. Instant messaging is a fairly standard offering, while very few vendors offer video- or audioconferencing internal to their software. However, groupware applications can be customized to fit an organization's needs. For example, e-mail can be either built into the groupware or software, or it can be designed to integrate with the company e-mail. Most groupware packages also allow for integration with external video- or audioconferencing software or systems.[37]

Holan suggests, "The easiest groupware systems to set up and require the least amount of user training are those that are hosted by the groupware vendor and run in the users' web browser. These usually involve the least amount of configuration, are the easiest to maintain, and the cost may be less since they can often be set up on a fee-per-use schedule." Low-end hosted applications can be as cheap as $10 per month. If considering a more expensive option, Holan suggests, "Always insist that a working demonstration be made available to your company for evaluation—before purchasing. And verify the support agreement offered within any contract—or the only collaboration you may get done is that of commiserating about how poorly the new software meets your needs."[38]

TIPS FOR USING GROUPWARE EFFECTIVELY

- **Choose a product that meets the organization's needs.** Because of the variability in features and cost, knowing exactly what you want the groupware product to do and how much you want to spend is important in making a good decision. Workgroup products can be configured to meet the needs of any industry, but there are some products created to target certain industries that may work just as well.
- **Provide training.** All of these products require at least some minimal training for the users. A trade-off for a less expensive groupware option, like ShareHQ Secure Workspaces, can be the additional training required to learn how to use a difficult-to-understand product.[39]

OTHER COMMUNICATION TECHNOLOGIES DESIGNED FOR GROUPS

A number of products designed for group work are readily available, including shared database systems, group calendars, groupware, project management software, and more. Each of these is a useful tool in certain situations with particular goals.

Electronic chat allows people in remote locations to have typed conversations with each other. A record of the conversation is generated as the comments made by chat participants appear on the screen. A chat room is an inexpensive way for distributed workers to engage in synchronous communication, making it an excellent tool when immediate feedback is required. Of course, keystroking comments can create lag time, which may be significant for people who are not good at the keyboard. While one participant prepares a long comment, others in the chat may have moved past that topic by the time the typing is completed. Although electronic chat can be a great benefit to remote workers, it can seem intrusive and disruptive if it is used too much, eroding some of the perceived benefit of telecommuting.

Shared databases make massive amounts of information easily available to all members of a virtual organization or team. Databases can store accumulated reference materials and be shared with team members or passed on to other teams. When well managed, these databases can ensure that the latest relevant information is accessible. They can provide a storage space for harvested information, so that each member of a virtual team has access to each member's knowledge. This is a very different form of knowledge management than would be found in a traditional, hierarchical organization.[40]

Group calendars allow team members access to each other's calendars, making it possible to schedule meetings and include deadlines. These packages offer some obvious advantages for scheduling but present some concerns for the manager of remote workers. When these packages are employed, the individual workers lose a little bit of control of their own schedules, and having more control over their schedule may be one of the reasons they chose to telecommute. Rules about who has the authority to put items on the schedule (and why) should be established.

Project management software allows team members to keep track of goals, tasks, and meetings. Some packages allow chats or polling and can be very useful to keep team members organized and coordinated.

Video windows are visual portals into an organization accessible on the web. To create a video window, a camera that is always on views a hallway or watercooler area in a company, and the live image is sent through the web. Employees located off-site can view the ongoing video whenever they want. These windows have been found to be particularly useful in helping remote workers feel connected to the centralized office.

Electronic Meeting Systems (EMS) have been used for a number of years in face-to-face situations. They range from simple polling mechanisms to software packages that allow everyone in the room working from a laptop to contribute to a display screen. The technology today allows for people to use electronic systems from remote locations. These packages offer a variety of tools designed to help accomplish common objectives of meetings, including brainstorming, analyzing issues, voting, outlining, and simply staying on task. Chat and video capabilities can be added to many of these systems.[41]

HIGH-BANDWIDTH LINES

The increasing availability of digital subscriber lines (DSL) and cable means that high-bandwidth lines will be more affordable, and those using the Internet for research will enjoy higher speeds than what is now available in much of the United States. Table 3-1 shows the number of Internet users with access to broadband from 1999, projected out to 2003.[42] If businesses decide to promote telework, then demand could substantially increase availability of broadband beyond what is projected here. In turn, increased availability of broadband may encourage more telecommuting.

The benefits of higher bandwidth for teleworkers were examined in one study in which teleworkers were surveyed before and after the installation of high-speed lines.[43] Participants in the study were asked to rate the effects of DSL on establishing communication, maintaining communication, gathering information, disseminating information, decision making, and efficiency, in terms of both their individual tasks and collaborative tasks. Results of the study indicated that DSL improved productivity across all areas.[44]

In addition to probably improving productivity, the increased availability of faster bandwidth will allow the use of more powerful collaboration tools. These tools may feature more than one window on the screen, one for the live image of the participants, one for data, and one for viewing video.

Table 3-1   Internet Users with Broadband, Projected to 2003

Year	Millions of Broadband Users	Millions of Internet Users	Percent of Users with Broadband
1999	2.50	58.00	4%
2000	5.51	75.75	7%
2001	10.98	90.00	12%
2002	17.77	103.88	17%
2003	25.09	118.25	21%

Source: Carl Van Horn and Duke Storen, "Telework: Coming of Age? Evaluating the Potential Benefits of Telework," *Telework and the New Workplace of the 21st Century,* 2000. Available: *http://www.dol.gov/asp/telwork/p1_1.html.*

FUTURE DEVELOPMENTS IN COMMUNICATION TECHNOLOGIES FOR TELECOMMUTERS[45]

- **Voiceover Internet protocol** allows a person to essentially make a phone call over the Internet. The voice quality is expected to improve to the point of rivaling ordinary telephone lines.
- **Unified messaging** brings together all the messages received through various media and converts them to a single medium. For example, you could retrieve both voice mail and computer-generated e-mail messages by telephone.
- **Wireless messaging** use is expected to increase, as the technology becomes less expensive, making telecommuters even more accessible. Already a wireless videoconferencing unit, the Tandberg 1000, makes videoconferencing possible from anywhere for a mere $5,490.[46]
- **Communicating over power lines** will soon be possible in some countries. Since power outlets are ubiquitous in first-world nations, this technology will allow Internet access from every room in the house. Unfortunately, because of the structure of the power grids used in the United States, the technology is not economical here. However, this technology may be important for globally distributed teams with members in Europe, South America, and Asia, where the service should be brought to market soon.[47]

DISCUSSION QUESTIONS

1. Of the communication technologies discussed in this chapter, which are asynchronous and which are synchronous?

2. What sort of e-mail policy do you think would be appropriate for telecommuters who tele-work only part-time? Would an e-mail policy be required for full-time telecommuters?

3. Should people ever use emoticons in professional communication? When would their use be appropriate and when not?

4. Think of your most recent job. What would be the minimum technological requirements to perform that job off-site?

5. Do you think higher bandwidth would improve employee job satisfaction? Why or why not?

ENDNOTES
1. Alison Overholt, "Virtually There?" *Fast Company* (March 2002): 108.
2. Overholt, 108–114.
3. Ibid.
4. Ibid.
5. Dana Hawkins, "Lawsuits Spur Rise in Employee Monitoring," *U.S. News & World Report,* 13 August 2001, 53.
6. Vital Statistics, "Big Brother Boss," *U.S. News & World Report,* 30 April 2001, 12.

7. James S. O'Rourke, "Cerner Corporation: A Stinging Office Memo Boomerangs," A communication case publication from the Eugene D. Fanning Center for Business Communication, Mendoza College of Business, University of Notre Dame, 2002.

8. Elizabeth Kelley, "Keys to Effective Virtual Global Teams," *Academy of Management Executive,* vol. 15 (2001): 132.

9. Carla Joinson, "Managing Virtual Teams," *HR Magazine* (June 2002): 69–73.

10. Kristen Bell De Tienne, *Guide to Electronic Communication* (Upper Saddle River, NJ: Prentice Hall, 2002).

11. David Sacash, "E-Mail and the CEO," *Industry Week,* vol. 250 (11 June 2002): 29–32.

12. Ibid.

13. T. Adams and N. Clark, *The Internet: Effective Online Communication* (Fort Worth, TX: Harcourt College Publishers, 2001).

14. Sacash.

15. Dave Johnson, "Smartphone Platform Wars," *Laptop* (July 2002): 32–46.

16. Michael A. Verespej, "The Old Workforce Won't Work," *IndustryWeek.com,* September 1998. Available: *http://www.industryweekcom.*

17. Deborah L. Duarte and Nancy T. Snyder, *Mastering Virtual Teams: Strategies, Tools, and Techniques That Succeed* (San Francisco: Jossey-Bass, 1999).

18. De Tienne.

19. Adams and Clark.

20. Timothy Captain, "Remote Access," *Laptop* (July 2002): 22–30.

21. Tonya Vinas, "Meetings Makeover," *eBusiness* (February 2002): 29–35.

22. Verespej.

23. Elizabeth Allen, personal communication with author, June 24, 2002.

24. Ibid.

25. Ibid.

26. Vinas.

27. Sarah Bean, "How to . . . Bring Rita & Rob Together," *Enterprise* (April/May 2002): 21–23.

28. Ibid.

29. Mike Byfield, "Personal Video Calling Is Ready for Prime Time," *Report/Newsmagazine* (Alberta Edition), vol. 29 (May 27, 2002): 40–41.

30. Martha Haywood, *Managing Virtual Teams* (Boston: Artech House, 1998).

31. Bean.

32. David Strom, "Web-based Discussion Forum Software," 2002. Available: *http://www.strom.com/places/wc.html.*

33. Overholt.

34. Ibid.

35. Ron Anderson, "Sharing Is Daring," *Network Computing,* vol. 13 (February 2002): 47–61.

36. Joanne Kelleher, "E-meetings Redefine Productivity," *Fortune,* 5 February 2001, S2–S12.

37. David Holan, personal communication with author, February 10, 2002.

38. Anderson.

39. Duarte and Snyder.

40. Ibid.

41. Carl E. Van Horn and Duke Storen, "Telework: Coming of Age? Evaluating the Potential Benefits of Telework," *Telework and the New Workplace of the 21st Century,* 2000. Available: *http://www.dol.gov/asp/telework/p1_1.htm.*

42. Ibid.

43. Adams and Clark.

44. Patricia Riley, Anu Mandavilli, and Rebecca Heino, "Observing the Impact of Communication and Information Technology on Net-Work," *Telework and the New Workplace of the 21st Century.* A dozen studies presented at a national symposium at Xavier University in New Orleans on October 10, 2000. Available: *http://www.dol.gov/asp/telework/p2_3.htm.*

45. Glenn Lovelace, "The Nuts and Bolts of Telework," *Telework and the New Workplace of the 21st Century,* 2000. Available: *http://www.dol.gov/asp/telework/p1_2.htm.*

46. Overholt.

47. W. Wayt Gibbs, "The Network in Every Room," *Scientific American,* vol. 286, no. 2 (February 2002): 38–53.

4 THE MEDIUM AND THE MESSAGE

B y the middle of the last century, sociologists and philosophers of popular culture began to examine the ways in which our society had changed. Following World War II, it was certainly more populous, less agrarian, more urban, faster paced, more industrial, and much more dependent on technology. Emerging new technologies, from the telephone to the automobile, from high-speed printing presses to television, had begun to shape our daily lives.

In 1964, a professor from the University of Toronto's Center for Culture and Technology declared that down through the ages, the means by which humankind communicates have determined our thoughts, our actions, and our lives. He further asserted that the mass media of the day were rapidly decentralizing modern living, turning the globe into a village and catapulting twentieth-century humans back to tribal life.[1]

It wasn't only the mass media that caught his attention. Marshall McLuhan focused on every available means by which people could exchange ideas. And, to the shock and consternation of many that followed the discussion, he asserted that the medium we select communicates as much as the content of our messages.

> *In a culture like ours, long accustomed to splitting and dividing all things as a means of control, it is sometimes a bit of a shock to be reminded that, in operational and practical fact, the medium is the message. This is merely to say that the personal and social consequences of any medium—that is, of any extension of ourselves—result from the new scale that is introduced into our affairs by each extension of ourselves or by any new technology.[2]*

McLuhan was arguing simply that the medium you select to deliver your message has an impact that may be equal to or greater than what you've chosen to say. His words have an oddly prophetic tone for those of us living forty years later in the twenty-first century.

WHY MEDIA CHOICE MATTERS

In the previous chapters, we have discussed the growth of telecommuting, the communication process, and communication technologies available for virtual work environments. In this chapter, we will look at the factors to consider when choosing the medium for your message.

With so many different media options available, choosing the right medium for a message can be complicated. But choosing the wrong medium can be a form of noise that will distort or disrupt the communication process. The medium itself has powerful symbolic meaning that communicates different aspects of the message, such as the level of formality and the priority of the information. If the medium creates expectations about the message that are subsequently violated, your communication objectives may not be met. For example, if you were to fire a member of your virtual organization, it simply wouldn't do to communicate this via e-mail. If your communication objective is simply to let the employee know that he or she is no longer a part of the team, then if the text of the message is clear and is understood by the employee, your objective has been met. If, however, your communication objective is to convey this information in a way that would be perceived as considerate and caring—or to protect the relationship between you—then you would fail. Similarly, sending a sympathy message by e-mail would be inappropriate because it is simply too casual, too informal. Sending a typed sympathy note would be too impersonal. In this case, the best choice of communication technology would be pen and ink.

Virtual Wisdom ▼

No single form of communication will bridge the physical and social distance of distributed workgroups. For both managers and workers, the trick is knowing when to use e-mail and when to use voice mail, when a teleconference will be effective, and when a personal meeting is necessary. Remote interaction requires a greater sensitivity to the strengths and limitations of different media: Face-to-face conversations are often best for creative brainstorming or sensitive negotiation, while e-mail may be quite effective for transmitting citation lists, student assignments, and other clearly defined pieces of information.

—Laurie Putnam, Manager of Information Design, Aspect Communications[3] ▲

The symbolic meaning of media will be established in part by the organizational culture and in part by the broader social culture. In some organizations, for example, low-priority messages may be sent by e-mail, while telephone calls are perceived as having a slightly higher priority. In other organizations, the telephone may have no greater priority than e-mail—it may be seen simply as a work-style choice. The vast majority of organizations do share a similar view of letters and memos, with letters appropriate for external communication and memos reserved for internal communication.

Individual groups of remote workers also produce their own communication norms. While at one time it was thought the different technologies would be used in certain ways with predictable effects (a kind of technological determinism), there is now evidence to support the suggestion that uses of communication technologies, like other aspects of communication, are socially constructed. This means that the norms for use of various technologies for different groups will emerge as the groups begin using those technologies. This also implies that some groups, even within the same organization, will have usage norms that are different from other groups, and individual group members will display some norms that are different from those of the group to which they belong.[4]

Consider, for example, the manner in which various CEOs use e-mail and how their respective organizations adapt to their preferences. "I use e-mail every day, 50 times a day," said Neal Rabin of Miramar Systems. "It's an ongoing conversation with a vast number of people inside and outside the company." At 9:00 a.m. on a recent morning, he had 89 messages in his in-box. Mark Hurd, Executive Vice President and COO of NCR Corporation's Teradata Division, prefers to communicate by voice mail. "I'm not a big e-mail fan," he states. Hurd receives twice as many voice mails as e-mails.[5]

MEDIA CONSIDERATIONS

Keep the following practical considerations in mind as you choose the medium for your messages to off-site workers or virtual team members.

AN IMMEDIATE NEED FOR FEEDBACK

A need for immediate feedback may be based on a desire to minimize the time required for an interaction, to allow for questions, or to provide an opportunity to confirm understanding. Feedback comes in two basic types: sequential and concurrent.[6] Sequential feedback occurs when the sender has paused, waiting for a reply. The receiver takes a turn as sender, asking for confirmation or reiteration, and then gives the floor back to the sender. Concurrent feedback, or back-channeling, occurs while the sender is delivering a message. This type of feedback includes expressions such as "oh" or "uh-huh" and nonverbal cues such as smiling and head-nodding.

Functions of feedback include the receiver's acknowledgment of understanding (such as a nod of the head), indication of a lack of understanding (a confused expression), and clarification or correction of the sender's message (a question, for example, delivered sequentially, often through interrupting the sender). In a face-to-face interaction, these functions of feedback contribute to both the effectiveness and the efficiency of communication by enabling the sender to recognize when (and if) the receiver understands the message.[7]

Immediate feedback allows a sender to adjust a message according to the reactions of the receiver and to use patterns that maximize communication efficiency. Senders, for example, can deliver part of a message, check for understanding, then deliver the next part of the message, and so forth. The time taken to communicate is reduced when immediate feedback is available.

If you need immediate feedback, some communication technologies will function better than others. Face-to-face meetings, phone calls, teleconferences, or videoconferences will afford both concurrent and sequential feedback of several sorts. Electronic chat will allow sequential feedback. People sometimes make the mistake of thinking e-mail is a

Virtual Wisdom ▼

Use at least two methods of communicating critical information and deadlines.
—Joe Wynne, Subject Matter Expert
gantthead.com[8] ▲

good choice when they require feedback quickly. But it's useful to remember that with e-mail, the receiver of the message has complete control over when the message is actually received (meaning opened and read). The receiver may, indeed, return feedback directly after receiving the message, but that may not be until days after it was sent.

A RECORD OF THE INTERACTION

Many media offer a record of interaction, making them good choices if you want to follow the thread of an interaction after the fact. One advantage of keeping a record is the creation of a shared group or corporate memory. As we noted in Chapter 2, individuals experiencing the same event may have different perceptions of it. Adding to these differences are those that result from memory reconstruction and communication failure; thus, it's not surprising that people often have very different memories of precisely the same event. Whenever people work in groups or teams, creating and maintaining mutual knowledge becomes a challenge. This can be especially important for remote workers who may not be regularly reminded through casual encounters in the hallway of what happened in a meeting or what the team decided. In addition to serving as a reminder for existing team members, new team members can gain a wealth of information about a group's history by reading archived records.

There are other times, however, when you may not want to keep track of everything that was communicated and who communicated what. For example, when your team is brainstorming, a temporary record of ideas is useful, but the intention to keep a permanent record that others might see later may inhibit some creative thinking. People may begin to judge their ideas before expressing them if they become concerned about their ideas being recorded (and attributed to them). All of this can be counterproductive for brainstorming.

PRIVACY

Many forms of communication may seem private, but are not. Earlier we noted that some employees believe the e-mail they use at work is private when, in fact, many companies closely monitor employee e-mail. Similarly, documents and e-mails that are sent with the intent of keeping them confidential sometimes do not remain that way. E-mail messages and attachments get forwarded to unintended recipients, and documents are turned over to unintended audiences. Think of how many times you've seen a document labeled "confidential" flashed across the television screen during a *20/20* or *60 Minutes* program, with passages highlighted. Whoever wrote those documents certainly didn't expect to see them on TV. The tobacco industry didn't expect to be forced to disclose documents kept secret, some for 40 years. Donald Garner, law professor at Southern Illinois University, remarked that the tobacco companies "are dying the death of a thousand paper cuts."[9] Even a reasonable and balanced message may have passages that seem sinister when taken out of context. So, as you choose a medium and construct a message, consider the potential secondary or shadow audiences. You may not have much control over who ultimately sees your message, because once you send it, the message is no longer yours. It belongs to the receiver (intended or not) to do with as he or she sees fit.

LEVEL OF FORMALITY AND PRIORITY

If the formality of the message doesn't match the formality of the medium, the message may be seen as inappropriate. Some oral or e-mail messages may require a follow-up letter as a formal record. For example, an employer will often formalize a job offer made and accepted by phone with a letter, and, in turn, a new employee will send a letter of acceptance after the verbal agreement. The level of formality can also depend, in part, on the context of the communication. Face-to-face communication can be considered informal when you chat with your boss in the elevator. However, if your boss schedules a meeting in his office, that face-to-face encounter will feel more formal.

Like level of formality, the level of priority (or urgency) of a message can be suggested by the medium. Of course, the content of the message itself can certainly convey priority, but the medium can contribute directly to this interpretation. The perceived priority of information delivered by different media can vary among organizations, since much of the perception is derived from the organizational culture. If your boss routinely sends you e-mail, then any e-mail from her may not be seen as a particularly high priority. She'll grab your attention, however, if, on a rare occasion, she leaves a phone message. And, with few exceptions, overnight deliveries from an express package service don't sit around unopened. They're almost always seen as urgent forms of communication.

PARTICIPATION REQUIRED

Certain activities and communication objectives require the active participation of others. Brainstorming, negotiation, and conflict mediation are just a few examples. Different communication technologies allow for varying forms and degrees of participation from others. In some situations, such as conflict mediation, face-to-face meetings would be preferred over just about any other technology, but in other instances, such as negotiation or brainstorming, any number of communication technologies would suffice. In fact, some researchers suggest that computer-mediated communication may improve outcomes in activities such as brainstorming.

When groups brainstorm, they are instructed to refrain from criticizing ideas, to produce as many ideas as possible, to improve or combine ideas already suggested, and to be creative.[10] The underlying tenet of brainstorming is that people will be able to produce more ideas and better ideas when they work as a group than when they work alone, because the thinking of each person is often stimulated by the ideas of the other people. While it sounds intuitive, the actual results of brainstorming sessions are often less than promising. In fact, research shows that brainstorming groups often perform less well than nominal groups (people working alone to produce output that is then combined). Many theories have been offered to explain this finding. Explanations include evaluation apprehension (people are afraid of having their ideas and themselves judged), free-riding or social-loafing (people on the team let the other team members do the work), and production-blocking (people are blocked from producing ideas because others are expressing ideas).[11]

These theories suggest possible advantages for brainstorming by computer-mediated communication. If participants are anonymous, which is possible with some software, then the effects of evaluation apprehension may be alleviated. Production-blocking would not be a problem with computer-mediated brainstorming because participants could read the responses of others and key in their own ideas whenever they wanted. In a study comparing the ideas of virtual brainstorming groups to nominal groups, researchers found that virtual groups produced fewer redundant ideas than nominal groups, although they did not produce more ideas or any better ideas than nominal groups.[12] Still, some users are believers. Bob Flynn, information services manager at one of GE's units, began using electronic brainstorming years ago. "This takes the mechanics out of arriving at a consensus," said Flynn. "We don't need flip charts, we don't need tape recorders, we don't need someone writing down their version of what's happening."[13]

EXPERIENCE WITH THE MEDIUM

It is not enough to provide remote workers with technological machines and applications. They must also be provided with appropriate training in how to use them. Even in face-to-face situations, lack of experience with equipment can cause annoying delays and disruptions to the work process. This is compounded in a virtual workplace, where difficulties with equipment may shut down communication completely.

Beyond the technical training, remote workers should also receive training in electronic collaboration. Research indicates that people experience higher levels of perceived mental load and frustration when using computer-mediated technology for a group task.[14] While this is undesirable, an even worse outcome is the inability for some groups to produce ways to share information and collaborate, even though they may be trying. In one study, despite repeated attempts, some groups failed to complete a task because group members were unable to figure out how to structure information in order to share it effectively in a computer-based

> **Virtual Wisdom ▼**
>
> *Companies often make the mistake of investing heavily in technology without making a corresponding investment in people, planning, processes, and training.*
>
> —Martha Haywood, Senior Consulting
> Partner at Management Strategies[16] ▲

format. Other groups that were able to successfully structure the information succeeded at the task.[15] This suggests that managers should consider the level of experience and training of remote workers when choosing a medium for complex tasks.

EMOTIONAL CONTENT OF THE MESSAGE

There are often times when the message itself is not the only communication objective. Negative emotions that might indicate a problem often don't emerge as easily or as clearly in a virtual environment. It is easier to detect and to fix problems that can cause negative emotional reactions with media that offers more nonverbal cues. As you nurture a developing relationship, or seek to resolve conflict, a synchronous medium may be your best choice.

On the other hand, research suggests that negative information may be conveyed with fewer distortions when using computer-mediated communication versus face-to-face communication. People ordinarily find it unpleasant to be the bearer of bad news, especially if it has personal consequences for the receiver. The temptation for many deliverers of bad news is to sugarcoat the message in ways that make it seem less negatively consequential to the recipient. The psychological unpleasantness of delivering bad news is relieved to some degree when the recipient of the news is not immediately visible or able to provide a candid, instantaneous reaction. As Stephanie Watts Sussman, assistant professor of information systems at Case Western Reserve University's School of Management, puts it, people are more comfortable delivering negative information if they "don't see the discomfort" it causes.[17] Computer-mediated communication may allow the deliverer of the message to be more honest and accurate.

MEDIA RICHNESS

Some theories of mediated communication focus on its drawbacks because of its limited capacity to transmit nonverbal cues. The main idea of these theories is that different media have different capacities for transmitting communication cues. Any media may be suitable for transmitting simple cues, but more complex and involved messages may require media with a higher capacity. Two of these theories are *media richness theory* and *social presence theory*. (For a thorough review of these and related theories, see "Wired Meetings," by Janet Fulk and Lori Collins-Jarvis.)[18]

Media richness refers to the ability of a medium to transfer information and create shared meaning in a given time interval, which depends on the amount of information and the number of channels through which information can be delivered with a given medium. Measures of media richness include the immediacy of feedback, the ability to communicate using multiple cues, the use of natural language rather than numbers, and the ability to readily convey feelings and emotions.

Virtual Wisdom ▼

Virtual teaming isn't something anyone planned. It happened because the technology was there.

—David Gould, virtual teams expert[19] ▲

Media richness theory suggests that outcomes for a task can be improved by matching the media richness to the type of task, known as the task-media fit hypothesis. The theory focuses on the information processing demands of a situation to evaluate the task. In situations in which there is high uncertainty about a task and the situation is equivocal (multiple meanings are possible), there is a need for high media richness. The theory suggests that the more sources of information (e.g., tone of voice, facial expression, and so on), the less likely it is that misunderstanding will occur with ambiguous or equivocal tasks. For unambiguous, low-equivocality tasks, leaner media are a better choice, because richer media would create inefficiency by providing too much unnecessary information (see Table 4.1).

Although media richness theory focuses on the effects of media use, early research into the theory looked only at media preferences. In one study, various communication tasks were described and people were asked to choose which media they would prefer to use to complete the task. While the results of this study supported media richness theory for choosing a medium, they did not offer any insight on the effects of actually using various media on the outcomes of a task.[20]

However, later studies did examine the effects on performance outcomes predicted by the task-media fit hypothesis. The good news for managers of virtual work groups is that while the media richness theory is supported by studies of preference, it isn't completely supported by studies of group performance. One study required participants to work together to solve different types of problems, including a negotiation problem that required conflict resolution. The participants communicated using face-to-face, videophone, telephone, or computer-mediated communication. The task-media fit hypothesis would predict that performance outcomes would be improved if the type of problem was a good fit with the media richness provided by the medium used. However, the results of the study showed that decision quality was no better for audio and video users than it was for face-to-face communicators, even for the very demanding negotiation problem. However, the decision quality was better for both audio and video communicators than for computer-mediated communicators.[21] For managers of remote workers, this means that a face-to-face meeting for a more complex problem may not be required to produce a good solution. Rather, a phone conference or a videoconference may be just as effective.

Of course, many remote workers depend primarily on e-mail and electronic chat for communicating. While some studies show a difference in outcome quality with computer-mediated

Table 4-1   The Task-Media Fit Hypothesis

Increasing potential richness required for task	Media for Group Communication			
	Increasing potential richness of information →			
Task Type(s)	**Computer Systems**	**Audio Systems**	**Video Systems**	**Face-to-Face Communications**
Generating ideas and plans	Good fit	Marginal fit; information too rich	Poor fit; information too rich	Poor fit; information too rich
Choosing correct answers: intellective tasks	Marginal fit; medium too constrained	Good fit	Good fit	Poor fit; information too rich
Choosing preferred answers: judgment tasks	Poor fit; medium too constrained	Good fit	Good fit	Marginal fit; information too rich
Negotiating conflicts of interests	Poor fit; medium too constrained	Poor fit; medium too constrained	Marginal fit; medium too constrained	Good fit

Source: Brian E. Mennecke, Joseph S. Valacich, and Bradley C. Wheeler, "The Effects of Media and Task on User Performance: A Test of the Task-Media Fit Hypothesis," *Group Decision and Negotiation,* vol. 9 (2000): 507–529.

communication and some do not, one consistent finding when comparing this medium to other richer forms of communication, such as video- or audioconferencing, is that it takes longer for people to reach agreement with computer-mediated communication. Even in studies where there are no significant differences in decision quality, consensus, or satisfaction, the time to get to a decision can be longer when computer-mediated communication is used.[22]

Despite research indicating that the quality of decision making through some electronic means of communication may be just as high as face-to-face decision making, the perception among the decision-makers is often that the decisions are of lower quality. A few studies of media richness that show no significant differences in the quality of the outcomes expected based on the richness of the communication media do show an impact on the satisfaction of the communicators and the desire for future interaction. Even people using audioconferencing or videoconferencing technologies for meetings may be less satisfied than those communicating face to face. This is especially true when the tasks to be accomplished are complex or are personally involving.[23]

These findings suggest that a good experience in group decision making and problem solving is made up of more than simply a good outcome. Managers of remote workers may choose to use richer media when there is more than just the quality of the outcome to consider, for example, when time is an issue or when there is a desire to ensure that communicators feel satisfied with the communication.

SOCIAL PRESENCE

Lower levels of satisfaction with the process of computer-mediated communication may be due to lower social presence of the participants. Social presence is the degree to which the combination of cues transmitted by a medium are able to suggest or represent the other person in an interaction. Social presence is related to the ability of a medium to transmit nonverbal cues, such as attentiveness, responsiveness, and other cues that are important in developing relationships. Social presence varies across media. Face-to-face meetings have the highest degree of social presence.[24]

Like media richness, the desired level of social presence in a medium depends on your goal. If your aim is simply to transfer information, a high degree of social presence is not required. However, for more interpersonally involving tasks, such as conflict resolution or negotiation, high social presence would be preferred.

This poses an important problem for managers of virtual work environments or leaders of virtual teams who are challenged with helping remote workers develop and maintain personal relationships. The need for relationships to form among distributed team members is among the major concerns of those who study the success of virtual teams. Virtual work groups that use text-based, computer-mediated communication often share less socially oriented communication than groups communicating face to face. Many argue that the technologies used most often by virtual teams lack media richness and social presence and are, therefore, unsuitable to the development of relationships among members. Furthermore, they argue that building relationships, particularly building trust, is vital for collaborative work.

THE VALUE OF FACE-TO-FACE MEETINGS

Meeting face to face provides the richest media and the highest level of social presence; therefore, it is good practice for remote workers to meet face to face for their first meeting[26] and to continue to do so periodically. These meetings provide an opportunity for members of organizations to play out political processes and validate social relationships. These aspects of meetings may be more difficult to engage when communication is electronically mediated. Since electronic communication is more likely to

Virtual Wisdom ▼

Videoconferencing is not necessarily a replacement, but a supplement or complement to travel.

—Karl Kebs, AT&T[25] ▲

be work-oriented, fewer side comments and less personal sharing arise than one might experience with a scheduled face-to-face meeting, or even an encounter at the coffeepot among co-located colleagues. This strictly business type of communication makes it difficult for distributed co-workers to develop either professional relationships or personal trust. While telecommuters attribute some of their increased productivity to their freedom *from* meetings, this loss of casual interaction may be an unrecognized cost in this.[27]

Managers in a virtual environment should understand the potential loss to employees that do not have face-to-face meetings but should also know that the need for rich media and social presence is not completely understood. Some researchers believe that theories of media richness and social presence do not take into account a number of factors that may affect the experience of electronically mediated communication. They believe that it would be naive to assume that the loss of one "unit of communication cue capacity" is equal to a loss of "one unit of understanding." Their reasons for this assertion can be summarized into a central theme: The difference in face-to-face communication and computer-mediated communication may be qualitative rather than quantitative. In other words, the amount of effective communication may be the same through either medium; it simply may be accomplished by different means.[28]

Managers of remote workers that are not able to meet face to face should also note that theories of media richness and social presence consider the effects of technology only in terms of restricting or filtering cues. They don't consider the ability of individual communicators and groups to adapt to different media when they recognize its limitations. For example, media richness theory suggests that a medium low in media richness would not be effective for a complex, uncertain task. However, the theory doesn't consider the responsiveness of the group trying to accomplish the task. Perhaps the group could restructure or modify the task to lower the uncertainty or reduce the complexity, thus changing the task to match the medium. Perhaps groups could even be trained to do this.[29]

A challenge for managers is to avoid viewing electronically mediated communication as simply representing a loss, but rather as a trade-off. While there is some loss in the reduction of communication cues, there are gains as well. For example, evidence suggests that audio- and videoconferencing can lead to more equal participation than face-to-face meetings. Another example is the written and nonverbal communication that occurs between co-located members of a virtual team during a conference call with distributed members. The co-located members can communicate with each other without the knowledge of the other participants in the conference call. That simply could not happen in a face-to-face meeting.[30]

FORMING IMPRESSIONS WITH COMPUTER-MEDIATED COMMUNICATION

The major concern with media richness and social presence for virtual work environments is that the reduction in cues will inhibit the ability of remote workers to form relationships, trust each other, and collaborate effectively. But not everyone agrees that these require rich media or high levels of social presence. Some theories suggest that people are typically motivated to form impressions of others right away and may then be motivated to develop relationships. Thus, they are driven to come up with ways to gain social information, despite the limitations of the available communication medium. Furthermore, it is suggested that developing relationships does indeed happen with computer-mediated communication; it's just that the process is slower than when face to face. In the short run, therefore, social development is incomplete. Impressions may form, but they are less detailed and more extreme.[31] The negative effects of low social presence are most likely to be found in short-term groups with no existing history and little expectation of future interaction. When more time is available and future interactions are likely, impressions and relationships do form.[32]

In fact, in certain circumstances those engaged in computer-mediated communication can experience intimacy and interpersonal assessments of their communication partners that exceed the ones of those engaged in analogous activities face to face.[33] This is explained in part by the tendency for people to form impressions of others based on their group memberships, if that is the only information known to them. For example, if you are part of a virtual team and the only thing you know about the other members of the group is that they are also on your team, you will form an impression based on that team membership. You also will be likely to overestimate the similarities of group members to each other and to yourself, since the only thing you know about these people is a team membership that you all have in common.[34] For some remote workers, being perceived as just another member of the group is positive thing. For example, disabilities that may have made workers stand out in a traditional work environment can become socially irrelevant in a virtual work environment. On the other hand, overestimating similarities can aggravate communication challenges by leading communicators to assume that others think and feel the same way they do and that their messages are understood.

When people do share social information, regardless of the medium, they usually attempt to present themselves as positively as possible. Computer-mediated communication offers some advantages in this regard in that it allows for the use of self-presentation strategies that wouldn't be available in face-to-face communication. Messages sent by computer can be well thought out, double-checked, and revised before they are sent. Unintended nonverbal messages, including physical appearance, are curtailed. The amount of thought required during the interaction is reduced as well, because there is no demand to attend to the nonverbal cues of the interaction partner, meaning that people can pay more attention to the preparation of their own message.[35]

For managers of virtual work environments, these theories suggest that forging relationships between remote workers may be less of a challenge over the long term. However, with a short-term commitment, as is often the case with virtual teams, forming relationships may be a significant challenge. A strategy that managers may use is providing visual information about the team members. One study of the virtual teams looked at the effects of the presence of photographs of the partners in computer-mediated communication on the impressions they formed of each other. The study showed that seeing photographs of partners promoted social attraction in short-term virtual groups.[36] For short-term commitments, providing visual information, for example, posting pictures of team members on a team web site, can be beneficial to creating interpersonal relationships—and developing trust—within the group.

COMMUNICATING TRUSTWORTHINESS

Why is building trust so important? Trust is recognized as an important element in successful collaboration for setting goals, creating shared values, and resolving conflict. All of which contribute to an organization's competitive advantage. Building trust is an important issue for managers of virtual work environments because nonverbal cues are an important part of the process.

Trust is a complex psychological construct that can be viewed as a two-dimensional concept, composed of the *perceived intentions* of others and the *perceived competencies* of others.[37]

We may have different types of trust in others that will shape and be shaped by the kind of relationship we have with them. We may have trust in someone's ability. (For example, we may trust that our dentist knows what she is doing.) We may have trust in someone else's regard for our welfare. (We may also believe that our dentist is concerned about what is best for our teeth.) We may have trust in someone's regard for privileged information. (As much as we trust our dentist in other areas, we may not tell her our marital woes. We would probably reserve that information for a friend or counselor.) We may have trust in someone's relational commitment. (We may trust that a friend will continue our friendship, even if we reveal information about ourselves that makes us feel vulnerable or information that we know the friend won't like.)

Trust is especially important in a virtual workplace for both managers and remote workers. Managers aren't able to monitor employees in the same way that they can when employees are centrally located. Traditional organizational hierarchies have clear roles and controls achieved through monitoring and directing. Virtual organizations are more laterally structured, and monitoring and supervising are not possible in the traditional sense. Trust must replace that control. Similarly, remote employees have to trust that they are being appropriately recognized and considered fairly for projects, and that their interests are of concern to their managers when they are not physically present to represent themselves. They want to know that senior management is committed to their projects and concerned for their needs.

> **Virtual Wisdom** ▼
>
> *If Big Bob looks over his Dilbertville and doesn't see cubicles filled with busy workers, he's going to wonder and feel uncomfortable about what he's paying the Little Bobs to do.*
>
> —Charles Grantham, President, Institute for the Study of Distributed Work[38] ▲

Eva Kasper-Fuehrer of Unisys in Austria and Neal Ashkanasy of The University of Queensland in Australia posit that trust is built on the communication of trustworthiness. They define communication of trustworthiness as

> *an interactive process that affects, monitors, and guides members' actions and attitudes in their interactions with one another, and that ultimately determines the level of trust that exists between them.*[39]

They argue that communication of trustworthiness depends on effective and dependable information and communication technology. Ineffective communication technology can disrupt the communication process and cause a rift in developing trust. According to these authors, information and communication technology can fail for any number of physical or human reasons. Physical failings include low bandwidth, lack of standardized equipment or compatible products, and system breakdowns. Human failings can include ineffective use of the technology for communicating trustworthiness (e.g., not responding to e-mails promptly).

As previously noted, face-to-face meetings early in the development of relationships within virtual organizations are particularly conducive to building trust, in part due to the nonverbal cues that are present. Of course, for many virtual organizations and distributed teams, meeting face to face simply isn't possible. In the absence of these nonverbal cues, people may use first impressions and stereotypes to extend a form of "swift trust" that is initiated quickly.

Once "swift trust" has been established, remote workers can help build on this by expressing excitement about a project, offering enthusiastic feedback after tasks have been assigned, completing assigned tasks, and delivering work on time. The past performance of remote workers becomes a means by which other team members may assess their trustworthiness. However, this creates a sort of trust that is particularly fragile because it deemphasizes the interpersonal aspects of trust. After trust is built in this way, it can be easily destroyed by behaviors that are inconsistent with the perceived role of the person in the team, such as missing deadlines or not responding to messages. These behaviors can make a once-trusted team member less trusted.[40]

Sirkka Jarvenpaa of the University of Texas and Dorothy Leidner of INSEAD conducted a study on the development of trust in global virtual teams. Based on their findings, they offer the following distinctions between communication behaviors of virtual teams that facilitate the development of trust and those that do not.[41]

- **Social communication.** Groups that share social information early on in the formation of the team develop higher levels of trust than groups that do not share social information. As groups with high trust progress, sharing social information, which is integrated into discussions of tasks, continues but does not substitute for progress on the task.
- **Communication conveying enthusiasm.** Teams with higher levels of trust express more enthusiasm and optimism. They express an excitement at being part of the team and offer encouragement to each other.
- **Coping with technical and task uncertainty.** Teams with higher levels of trust have more communication about when they will be available to work and about goals and tasks. Teams with lower levels of trust have little clear communication about goals and tasks and procedures for dealing with uncertainty over the task or with technical problems.
- **Individual initiatives.** Teams with lower levels of trust have less individual initiative. They are less self-directed and want to receive instructions on what to do and leave decision making to others. Their communication reveals a hesitancy to commit (e.g., I *think* I can).
- **Predictable communication.** Trust is more forthcoming and more easily maintained when teams have predictable communication patterns. Team members that create a failing team go away on vacation without forewarning the others. They don't respond to messages, leaving the rest of the team confused and concerned. This inhibits trust. Members of teams with higher levels of trust offer quick responses to messages, verifying that messages have been received and read.
- **Leadership.** Leaders of teams with higher levels of trust emerge after exhibiting some skills that indicate rightness for the job, but leadership is shared among members, depending on the task. Leaders of teams with low levels of trust appoint leaders based on whoever had communicated first or sent the most messages in the early stages of the team formation. These may not be effective leaders. They may use a punishing approach, in the form of complaining comments, to lead the team, rather than positive reinforcement, or praising comments.

Kasper-Fuehrer and Ashkanasy identify two other factors important in the development of trust in organizations. The first is a common business understanding, which refers to the agreed-upon understanding among members of a virtual organization or virtual team of their organizational identity, what they stand for, and what they do. Members must agree upon organizational expectations and goals. Of course, these are important in a traditional organization as well. The difference with a virtual organization is not in their importance, but in how they are communicated. In a traditional organization, such things as artifacts (furniture, office décor, location, and so on), the structure of the organizational chart, and implicit organizational culture communicate to employees about organizational identity and goals. These need to be explicitly discussed and agreed upon in a virtual organization. These discussions and agreements may take the form of a clarification of the tasks and roles for each employee or team member.[42]

The second factor noted by Kasper-Fuehrer and Ashkanasy is business ethics. Business ethics has an obvious and intrinsic association with trust. While expected ethical business behavior is often communicated informally in traditional organizations through organizational culture and norms, it may need to be communicated more formally through a policy on business ethics in the case of a virtual organization. Organizations that have a reputation for ethical behavior will have their behavior as an illustration of their policy.[43]

The very communication technologies that make virtual work environments possible also make them challenging for managers and remote workers. Each communication technology offers advantages and disadvantages. For example, e-mail, the primary communication tool for most remote workers, is convenient, easy to use, and inexpensive, yet it is asynchronous and void of nonverbal cues. The challenge for managers and remote workers is to select from the numerous communication technologies available the one that is appropriate for the communication objective. No single medium is always the best choice.

In this chapter we have discussed a number of media considerations for remote employees, virtual team members, and managers in virtual organizations. The need for immediate feedback, privacy, formality of the message, and the desire to form relationships all must be taken into account. McLuhan's words on the importance of the medium are particularly applicable to the virtual work environment. The medium is not the entire message, but it is clear that the medium will influence how messages are interpreted and what their ultimate effectiveness will be in reaching communication objectives. In a virtual work environment, even reaching long-term organizational goals depends on strategic medium choice.

DISCUSSION QUESTIONS

1. Discuss the symbolism of various media in your experience. How do companies use media to indicate formality? How do they signal a more personal message?

2. What media should organizations use to send collection messages?

3. What type of media would be best for communicating messages with emotional content? Is it best for emotional messages to be delivered with verbal and nonverbal cues? With opportunities for immediate feedback?

4. Discuss your thoughts on impression formation, social presence, and media richness theory. Have you ever met someone after exchanging e-mail messages for some time? Was that person as you had expected?

5. Do you think it would be difficult to develop trust in virtual team members?

ENDNOTES

1. Marshall McLuhan, *Understanding Media: The Extensions of Man.* (New York: The New American Library, 1964).
2. Ibid., 23.
3. Laurie Putnam, "Distance Teamwork," *Online,* vol. 25 (2001): 54–58.
4. Tom Postmes, Russell Spears, and Martin Lea, "The Formation of Group Norms in Computer-Mediated Communication," *Human Communication Research,* vol. 26 (2000): 341–371.
5. David Sacash, "E-Mail and the CEO," *Industry Week,* vol. 250 (11 June 2002): 29–32.
6. R. M. Kraus and W. Weinheimer, "Concurrent Feedback, Confirmation, and the Encoding of Referents in Verbal Communication," *Journal of Personality and Social Psychology,* vol. 4 (1966): 343–346.
7. Alan R. Dennis and Susan T. Kinney, "Testing Media Richness Theory in the New Media: The Effects of Cues, Feedback, and Task Equivocality," *Information Systems Research,* vol. 9 (September 1998): 256–275.
8. Joe Wynne, "The Care and Feeding of Virtual Teams," 2000. Available: *http://myplanview.com/expert10.asp.*
9. Myron Levin, "Years of Immunity and Arrogance up in Smoke: Tobacco Companies Face 'Death of a Thousand Paper Cuts' from 40 Years Worth of Incriminating Documents," *Los Angeles Times,* 5 October 1998. Excerpts available at *http://ash.org.*
10. Alex F. Osborn, *Applied Imagination* (New York: Scribner, 1957).
11. Rene Ziegler, Michael Diehl, and Gavin Zijlstra, "Idea Production in Nominal and Virtual Groups: Does Computer-Mediated Communication Improve Group Brainstorming?" *Group Processes and Intergroup Relations,* vol. 3 (2000): 141–158.
12. Ibid.
13. Alice LaPlante, "90s Style Brainstorming," *Forbes,* vol. 152 (October 1993): 44–49.
14. Reze Barkhi, Varghese S. Jacob, and Hasan Pirkul, "An Experimental Analysis of Face-to-Face versus Computer-Mediated Communication Channels," *Group Decision and Negotiation,* vol. 8 (1999): 325–347.
15. Kenneth A. Graetz, Edward S. Boyle, Charles E. Kimball, Pamela Thompson, and Julie L. Garlouch, "Information Sharing in Face-to-Face Teleconferencing and Electronic Chat," *Small Group Research,* vol. 29 (1998): 714–744.
16. Martha Haywood, *Managing Virtual Teams: Practical Techniques for High-Technology Project Managers* (Boston: Artech House, 1998).
17. Stephanie Sussman and Lee Sproull, "Straight Talk: Delivering Bad News through Electronic Communication," *Information Systems Research,* vol. 10 (June 1999): 150–167.
18. Janet Fulk and Lori Collins-Jarvis, "Wired Meetings," in *The New Handbook of Organizational Communication,* edited by Fredric M. Jablin and Linda Putnam (Thousand Oaks, CA: Sage, 2001), 624–663.
19. David Gould, "Virtual Organization," 2000. Available: *http://www.seanet.com/~daveg.*
20. R. L. Daft, R. H. Lengel, and L. Trevino, "Message Equivocality, Media Selection, Manager Performance," *MIS Quarterly,* vol. 11 (1987): 355–366.
21. Brian E. Mennecke, Joseph S. Valacich, and Bradley C. Wheeler, "The Effects of Media and Task on User Performance: A Test of the Task-Media Fit Hypothesis," *Group Decision and Negotiation,* vol. 9 (2000): 507–529.
22. Dennis and Kinney, 256–275.
23. Jill M. Purdy, "The Impact of Communication Media on Negotiation Outcomes," *International Journal of Conflict Management,* vol. 11 (2000): 162–184.
24. J. Short, E. Williams, and B. Christie, *The Social Psychology of Telecommunications* (London: Wiley, 1976).

25. "Closer Than You Think," Conferencing Roundtable, *Communication News,* vol. 39 (March 2002): 12–17.
26. John Grundy, "Trust in Virtual Teams," *Harvard Business Review,* vol. 76 (1998): 180.
27. Fulk and Collins-Jarvis.
28. Ibid.
29. Ibid.
30. Ibid.
31. Jeffrey T. Hancock and Philip J. Dunham, "Impression Formation in Computer-Mediated Communication Revisited," *Communication Research,* vol. 28 (2001): 325–347.
32. Joseph B. Walther, Celeste L. Slovacek, and Lisa C. Tidwell, "Is a Picture Worth a Thousand Words?" *Communication Research,* vol. 28 (2001): 105–134.
33. Ibid.
34. Russell Spears and Lea Martin, "Panacea or Panopticon?" *Communication Research,* vol. 21 (1994): 427–460.
35. Walther, Slovacek, and Tidwell.
36. Ibid.
37. D. M. Rouseau, S. B. Sitkin, R. S. Burt, and C. Camerer, "Not So Different After All: A Cross-Discipline View of Trust," *Academy of Management Review,* vol. 23 (1998): 393–404.
38. Susan J. Wells, "Making Telecommuting Work," *HR Magazine,* vol. 46 (2001): 34–46.
39. Eva C. Kasper-Fuehrer and Neal M. Ashkanasy, "Communicating Trustworthiness and Building Trust in Interorganizational Virtual Teams," *Journal of Management,* vol. 27 (2001): 238.
40. Ibid.
41. Sirkka L. Jarvenpaa and Dorothy E. Leidner, "Communication and Trust in Global Virtual Teams," *Organization Science,* vol. 10 (1999): 791–816.
42. Kasper-Fuehrer and Ashkanasy.
43. Ibid.

5 MANAGING VIRTUAL WORK ENVIRONMENTS

In Chapter 1, we explored numerous advantages that virtual work environments offer both employees and organizations. Employees are able to save on commute time and enjoy more flexible schedules. Organizations are able to reduce real estate costs, increase productivity, achieve higher profits, improve customer service, gain access to global markets, attract better workers, and improve environmental conditions—all through the use of telecommuting.[1]

We also noted in Chapter 1 that actual levels of telecommuting are far below predictions made several years ago. We then discussed reasons why managers and employees might have reservations about off-site workers. Managers have a number of concerns about effectively managing remote workers without being able to see them. In addition, middle managers feel as if their own jobs are in jeopardy if those whom they manage—the very reason for their employment—can work without supervision.[2] Some managers also believe that nontelecommuting employees may resent employees who are permitted to work from home, which could negatively affect culture.[3]

Potential telecommuters see being out of the office as translating into being out of the loop. They feel that telecommuters are less connected to and less visible in their organizations, making them less likely to be considered for promotions and choice assignments. An even greater fear in tough economic times is that they would be most vulnerable to layoffs.[4] For some people, isolation and lack of social connection is a concern.

Still, numerous organizations, such as Merrill Lynch, IBM, and Georgia Power, have very successful telecommuting programs.[5] Many organizations with successful programs have more telecommuter applicants than openings. The difference in these organizations is a formal telecommuting program that involves planning, policy creation, and training for employees and managers.[6]

Many organizations reporting that they don't offer telecommuting actually allow informal, occasional telecommuting. However, these informal programs can cost three to five times as much as formal telecommuting programs. Costs can escalate and performance can suffer because there is no training, no framework, and no strategy. When this happens, managers are likely to blame telecommuting, although the real problem is poor planning.[7] In addition, informal programs don't offer the recruiting advantages of formal programs. You can't attract a candidate with telecommuting opportunities if you don't formally offer them.

Creating formal telecommuting programs is a better practice, but it means change—change for workers who must be self-directed, and change for managers who must trust workers to direct themselves. Supervising remote workers requires managers to shift their focus from time and attendance to results. They must shift their role from enforcer to facilitator. They must shift their leadership style from directive to transformational. Instituting telework without changing traditional styles of management can lead to trouble and eventual failure.[8]

MANAGING OFF-SITE EMPLOYEES

Managing remote workers offers a unique set of challenges. According to Wayne Cascio, Professor of Management at the University of Colorado at Denver, the most effective managers of virtual organizations have positive attitudes. They are optimistic about the success of virtual work arrangements and focus on solving problems rather than using the problems as a reason to give up. They are also results-oriented. Focusing on tasks and monitoring how time is spent are ineffective management strategies with remote workers. Effective telecommuting managers have good communication skills and know how to delegate and follow up.[9] Creative, flexible, innovative managers are better able to adapt to managing remote workers.[10] Remote workers fare best with managers who empower their employees, who offer regular feedback, and who are good facilitators.[11]

Successful management of remote workers doesn't just happen. Merrill Lynch spent four years studying how to best implement telecommuting before doing so in 1996. Their research resulted in a 21-page manager's guide to nontraditional work arrangements, training (which includes a workshop where managers and employees negotiate on issues such as measuring productivity that could negatively affect the telecommuting relationship), and a successful telecommuting program with over 3,500 employees currently working from home part of the week.[12]

One way that all managers can become more comfortable with the concept of virtual work environments and more effective at managing remote workers is to learn more about the associated challenges and how to deal with them effectively. Implementing and managing successful virtual work arrangements takes careful planning and training, but the rewards for both organizations and employees are well worth the effort. In this chapter, we offer a review of the best practices for managing remote workers and suggestions for their implementation.

> **Virtual Wisdom ▼**
>
> *People realize that a work-at-home arrangement is only as good as they make it. If production and performance suffer, they know it'll affect the whole outcome for both parties.*
> —Sean Scott, CIO, Womble Carlyle[13] ▲

THE JOB

Some jobs can be completed off-site, while others cannot.[14] The first step in determining whether a job can be completed from a remote location is to analyze the job itself. A traditional job analysis produces a description of the duties involved and the qualifications required to complete them. With this in hand, a manager can better assess whether all or part of that job can be completed off-site.[15]

Some aspects of the job to consider include[16]

- How much time a worker must spend in direct contact with customers, clients, and other employees;
- Whether completion of a job is in some way time-constrained;
- Whether sensitive information is required to complete the job that cannot be securely accessed or stored outside the office;
- Whether the information required for the job is truly portable;
- Whether the outcome of the work is dependent on location; and
- Whether the employee may need to be reached immediately.

Some jobs do not lend themselves well to telecommuting. Work that requires access to expensive equipment or large machinery wouldn't be suitable. But most knowledge jobs can be completed all or in part off-site. Some jobs can be done permanently off-site, while other jobs are suited for telecommuting only during certain seasons, times, or phases of the work. For example, a human resource worker, who needs to be on-site regularly so that people may stop by and ask questions or discuss problems, may work from home when performing information-based tasks, such as developing training courses or updating policy manuals.[17]

Virtual Wisdom ▼

It's not so much a question of "Is this job or functional area suited to telecommuting or not?" The question is, "Are there one to three days' worth of work in this job that can be done as well or better from home?"

—Gil Gordon, consultant[18] ▲

Tasks that require individual work, a great deal of concentration, very little face-to-face interaction, and which produce measurable outputs—so that productivity can be assessed in terms of results rather than time—are best for telework.[19] An important goal for managers who want to implement a telework program is to create job descriptions that state whether a job is available for telecommuting.[20]

SELECTION OF TELEWORKERS

Teleworkers who are hired from outside the organization miss much of the valuable socialization process that traditional employees go through when they begin work. This socialization process introduces them to the culture of the organization, including the organization's norms, standards, and values. This process gives both the employee and the organization time to discover if there is a good fit between them. To compensate for the loss of this natural socialization, some organizations require their newly hired telecommuters to attend training on the internal workings of the organization. Silvia Orian, telecommuting sales representative for Xerox, was hired from outside the company. She had to spend two and a half weeks at a training facility before she could begin working from home.[21] Others assign mentors to teleworkers or create virtual teams that combine new hires and experienced workers.[22]

Another way to approach the socialization problem is to hire telecommuters from within the organization. Existing employees already know the way things are done in an organization and why. They know how long tasks are supposed to take and the level of performance that is expected.[24] They have already established that a match exists between their values and those of the organization.

Hiring teleworkers from within also allows managers to evaluate the work habits of tele-workers before assigning them to remote work. Teleworkers should have good attendance, solid relationships within the organization, and be top performers.[25] The best remote workers are independent, self-motivated, well organized, and need very little direct supervision.[26] Time management skills are important for employees to set and meet deadlines and to separate work from other responsibilities.[27] Good communication skills are also essential.[28]

Even with these skills and characteristics in place, telework is not for everyone. Different personalities and work styles can affect how well a person adapts to working off-site. Merrill Lynch employees who are considering telecommuting must first work for two weeks in a simulation lab, where they have no face-to-face contact with their manager. They are able to experience what it's like to communicate with their boss only by telephone and e-mail. They also learn how to manage minor problems with their computers, software, and other equipment. In this way, they can make an informed decision about whether telecommuting is right for them.[29]

Of course, in some situations, working off-site is a requirement, not a choice. That's the case in a completely virtual organization or team where the remote aspect of the work is understood and the issues of selection involve questions about fit with the rest of the team or organization. According to Martha Haywood, Senior Consulting Partner at Management Strategies, a consulting firm that specializes in the management of distributed teams, the ideal remote worker is a good fit with the team or organization in terms of goals, process, tools, and skills.[30]

- **Goals.** A good fit for goals means that employees and organizations share a common view of what is important. Having truly common goals requires a shared meaning of what those goals represent. For example, an organization may have the goal of "excellent customer service" but the meaning of "excellent customer service" won't be the same for everyone. Assessing fit in this area requires determining candidates' goals and what those goals mean to them, how their goals are prioritized, and how committed they are to achieving them. Examining a candidate's accomplishments and asking questions about how a candidate makes decisions can also be helpful in assessing fit for goals.

- **Process.** People and groups go about doing things differently. One group may spend a lot of time on the details of a project, while another may focus on getting as many projects done as quickly as possible. Organizational culture often communicates to new employees "how things are done" or the process expected, but in a virtual organization, culture is more difficult to communicate. A newly hired, remote worker won't know if a particular task usually takes one hour or four hours. Assessing fit for process before hiring can be useful in making the hiring decision or planning for communicating process expectations. Asking specific questions about how the candidate has done things in the past and direct questions about how the employee likes to work can help determine fit for process.

- **Tools.** A good fit for tools means that the candidate has the necessary equipment available to do the work and maintain communication and that the tools are compatible with those used by others in the organization. For remote workers, necessary tools also include adequate work-space, an extra phone line, and comfortable office furniture.

  Many organizations outfit their employees with the desired tools—including the office furniture, while other organizations want to determine if a candidate already has the necessary and compatible tools.

- **Skills.** In virtual work environments, some level of technical skills is desirable. Employees that are more comfortable and familiar with the equipment they're using are more likely to be able to deal with small problems when they arise. This can be particularly important for a distributed workforce, since there are no IT people in the next office to call for help. Simple equipment problems can lead to periods of downtime for remote employees who have no skills in this area. In addition to computer skills, remote workers need good communication, self-management, and collaboration skills. The answers to behavior-based questions, which ask candidates to give examples of a particular type of situation and how they handled it, can show how candidates have demonstrated their skills.

Finding a perfect fit may be impossible, but finding a comfortable fit and having a plan for creating alignment from that point is acceptable. Of these aspects of fit, tools and skills may be the easiest to align. Training and clear communication from management, however, can create alignment in goals and process, as well.[31]

Will Pape, cofounder of VeriFone, Inc., spent a lot of time considering the characteristics of the best candidates for virtual work environments. He cites strong communication skills as being of primary importance. He also notes the importance in taking the initiative in communication. "Staff members who wait to be asked, who don't take the initiative to inform, are going to slow productivity," asserts Pape. "You want people who won't hesitate to make that phone call or send that e-mail." Pape recommends evaluating a candidate's loyalty, work ethic, and ability to solve problems independently. He also notes the importance of a sense of humor. "People with a good sense of humor—who consequently are slow to anger—tend to deal more effectively with the frustration of the virtual workplace."

Virtual Wisdom ▼

It was an interesting time of my life, but [I] developed some very bad personal habits as a result that drove my wife crazy.

—Eric Cloninger,
former telecommuter[32] ▲

Pape acknowledges that not every successful virtual worker has each characteristic and that training and company practices can help with shortfalls.[33]

Remote workers and virtual team members also need to possess certain competencies to be effective. In their book *Mastering Virtual Teams,* Deborah Duarte and Nancy Snyder outline six competencies that virtual team members should possess.[34] They include the following areas:

- **Project management.** Good project management skills mean that a person can plan, coordinate, cooperate, collaborate, schedule, and deliver on time. Good project managers are able to record and share information effectively.
- **Networking.** Networking skills are social skills that enable a person to recognize stakeholders, take another's perspective, and interact effectively with people from inside and outside the organization. These skills also include choosing the best medium for communication in order to accomplish a goal.
- **Technology.** Skills in technology include an awareness of available technologies and the knowledge of how to use them as well as the ability to determine when the use of various technologies is appropriate.
- **Self-management.** Skills in self-management include the ability to prioritize, set limits, and learn and grow.
- **Culture.** Recognizing the importance of cultural differences and understanding their impact are valuable skills for virtual team members. Even more important is the ability to turn differences into advantages.
- **Interpersonal awareness.** Interpersonal awareness is the knowledge of oneself and how one's own behavior affects others. An important part of being interpersonally aware is recognizing areas where development may be needed.

Duarte and Snyder recommend that levels of competency be assessed in these areas and that development plans be created. Development efforts can include training, chats about best practices, mentoring, or assigned readings. Development should focus on both individual and organizational or team needs.[35]

No matter how good of a fit a person may be for a virtual team or how high the level of competence a person may have, one final consideration may override all others. That is simply that some people need more face-to-face time than others, and they may feel lonely if they work from a remote location.

EQUIPMENT AND LOCATION

When you select remote workers, you must first consider where they will actually perform their work and what equipment they will need. If your employees are to work from home, you should keep some things in mind. To be effective, teleworkers should have a dedicated space in which to work and store materials that is both comfortable and private.[36]

Initial equipment purchases to outfit a home office can be expensive. You should avoid the temptation to go with the least expensive equipment, however, or to place older units no longer used in the office in the homes of teleworkers. A technical failure that requires a visit to a telecommuter's home or that requires a telecommuter to unplug a unit and bring it in for repair can wind up being much more expensive than buying the very best equipment to start with.[37]

Applications that worked quickly in the office may be quite slow from home with a modem and standard phone line. You can address this problem by installing a broadband option, which is available in most areas for a monthly fee that is about the same as the cost of utilities for a single office.[38]

Security is another area of concern for people working from home. Telecommuters should be trained on the importance of security and the risks involved with lax practices. Employees dialing into the corporate network can spread viruses. Virus scanners and firewalls thus become important security tools. Michael Demaria of *Network Computing* suggests that organizations establish a policy stating that disabling a firewall or virus scanner would be grounds for termination.[40] Another suggestion is that the business computer in the home be reserved strictly for business and that another computer be available for personal use, if necessary. This can protect the corporate LAN from bugs that may infect an employee's computer during personal use.[41]

> **Virtual Wisdom ▼**
>
> *Back up your folks with a corporate help desk to walk them through the problems they're certainly going to encounter.*
>
> —William Pape, cofounder
> VeriFone, Inc.[39] ▲

Safety can also be an issue for those working at home. You must make telecommuters aware of safety standards and make sure that equipment works properly and is correctly installed to prevent problems. Requiring periodic safety inspections is also not a bad idea.[42] Providing employees with quality office furnishings is another preventive measure to reduce safety concerns.[43]

SUPERVISION OF TELEWORKERS

As in a traditional office, the supervision of employees involves establishing goals, setting deadlines, and guiding work efforts. With remote workers, however, this supervision involves little or no face-to-face time. Clearly communicating goals and monitoring performance without being able to physically observe employees is an added challenge for managers in virtual work environments. This can be particularly difficult for managers who have become accustomed to managing by walking around. Strategies for meeting these challenges include creating a telecommuting policy, establishing communication agreements, and building trust.

> **Virtual Wisdom ▼**
>
> *All you can do is manage products and output: you can't manage the process.*
>
> —Edward Lawler III, Director of the Center
> for Effective Organizations[44] ▲

TELECOMMUTING POLICY

Creating a formal telecommuting policy is a good way to manage expectations and prevent problems. A telecommuting policy should address issues such as hours, equipment provisions, technical support, security, hidden as well as anticipated expenses, sick leave, safety requirements, injuries, workers' compensation, zoning, and tax issues.[45] Having a policy of this sort provides a frame of reference, both for teleworkers and managers.

- **Hours.** Flexibility is one of the appealing features of telecommuting. Still, you may want to insist that telecommuters be accessible during certain times, perhaps for a required teleconference or to take client calls. In some cases it might be important that a remote worker be immediately accessible (if only electronically) during set home office work hours.
- **Equipment, technical support, and security.** You should draft and sign formal agreements about what sort of equipment will be provided, who will install it, and how it may be used. Having a procedure in place for equipment failure and technical support is also important. Failing to provide technical support is one of the most common mistakes for managers of remote workers.[46] Silvia Orian of Xerox identifies IT problems as her biggest struggle. If she has hardware failure, it can be weeks before someone comes out for service, and during that time she can do very little.[47] Your policy should include requirements for password protection, firewalls, virus scanners, and any other security tools you think are important. The policy should also specify how company-owned equipment would be returned after employment is terminated.
- **Hidden and unhidden expenses.** Expenses such as utilities, additional phone lines and phone-related services, broadband access, equipment maintenance and insurance, office furnishings and supplies, copying costs, postage, and costs to travel to the main office should be discussed. The reimbursable and nonreimbursable expenses should be clearly defined.[48]
- **Sick leave.** Telecommuters often develop a feeling of guilt for not working when they are sick. Even if they have a cold or flu that is serious enough to keep them home from a traditional office, they may not feel that they are too ill to sit in front of their computer. Therefore, they wind up working when they are sick. This may negatively affect the quality of their work and their attitude.[49]
- **Safety requirements.** Your policy should advise employees about the safety and environmental requirements that may be needed to protect them and the equipment they are using. Safety inspections should be discussed and the consequences of failing an inspection should be made known. You should be clear, for example, about who will pay for upgrades if the wiring in the employee's home fails inspection.[50]
- **Injuries, workers' compensation, zoning, and tax issues.** Companies should seek help from their legal counsel in creating the portion of their telecommuting policy addressing employment law-related issues. Similarly, employees should consult with their tax professional about the tax consequences of telecommuting.

Policies that address these issues can help managers and telecommuters feel more comfortable and manage their expectations. If you're still apprehensive about virtual workplaces, consider including a policy statement regarding who may be qualified to telework. Such statements may help to increase the confidence of managers and remote workers alike. Many companies limit teleworking opportunities to their top performers. Managers are simply more trusting of employees with a proven track record. In case things don't work out, however, you may wish to include a statement that telecommuting can be terminated due to poor performance.[51]

COMMUNICATION AGREEMENTS

Not all situations afford you the luxury of remotely managing only proven employees. Whether your remote workers are time-tested or not, establishing policies about communication will prove vital for your effectiveness and that of your employees. Feedback rules, for example, are important for off-site workers. If an e-mail or voice mail doesn't receive a response, we are left to only wonder why.

Consider this set of events: I sent an e-mail to a woman whom I didn't know at all about a project we had discussed. I had some concerns, some ideas, and some questions. I had spoken to her in person a few days earlier and found that she was very outgoing and responsive, so when I didn't get a response to my e-mail, it seemed inconsistent with my first impression of her. I wondered if she was out of town, or if she had decided that she didn't want me for the job, or if she thought my ideas and questions were stupid and didn't deserve a response. Then, I debated about sending another e-mail because I didn't want to seem pushy and demanding in the event that she had received my e-mail but just hadn't gotten around to responding to it yet. It was very uncomfortable. As it turned out, she had just been very busy and did finally contact me. She liked my ideas and I was given the job, but my imagination certainly had worked overtime while I was waiting to hear from her.

Gaps in communication (e.g., unanswered e-mail) inhibit the performance of remote collaborators and erode trust.[52] Rules should be established for changing outgoing voice mail messages and using an automatic e-mail response when a person is unavailable to respond to messages for more than a day. Norms for how long it should take to respond, even if that response is something to the effect of "I don't have an answer for you yet," should be established and explicitly communicated. Other communication agreements should reflect the following guidelines:

- **Regular meetings.** Set up regularly scheduled face-to-face or electronic meetings.[53] This may seem like a counterintuitive suggestion, since one of the clear appeals of working from a remote location is escaping from meetings. But weekly meetings can go a long way toward building trust, cultivating relationships, strengthening culture, and improving performance within a group. Information sharing in regular meetings can help remote workers remain aware of what others in the organization are doing. It can also invite feedback and constructive criticism about performance and provide a learning opportunity about what might have been done differently to improve the outcomes from the week.[54] Research shows that regular meetings are critical to the stability and functionality of virtual teams.[55]

- **Communication technologies.** Using a variety of communication technologies will prevent boredom and communication burnout. E-mail may be the primary medium for communication in distributed teams, but synchronous communication should also be used regularly. Lisa Kimball and Amy Eunice of Caucus Systems, Inc., recommend using a variety of technologies and creating norms for their use. For example, with e-mail they recommend establishing norms for response times, forwarding e-mails, and sending blind copies. Norms can be established for how the receipt of a message will be acknowledged, how others will be advised if a remote worker is unavailable for a few days, and so forth.[56]

- **Recreate context.** For remote workers the context of communication is removed. Setting agendas for meetings, having a clear goal, sharing background materials in advance, having regular meetings with the same participants, and having clear roles and responsibilities can compensate for the loss of context.[57]

- **Prioritize messages.** In her work with virtual teams, Martha Haywood has found that message senders on teams with successful communication take responsibility for prioritizing their e-mails. She recommends creating a subject line code that will indicate priority and tell recipients whether the message is intended for them or if they are just being copied on the message.[58]

Effective communication is a critical success factor for virtual teams. The most important challenge is maintaining continuous communication. Planning communication processes in advance is a preventive measure that can contribute significantly to the success of virtual organizations.[59]

BUILDING TRUST

One of the causes for managers' hesitancy to support telecommuting is their perceived loss of control over employees' behavior. In the traditional office, managers are able to use several categories of control strategies over employees. They can use behavioral control methods, such as close supervision and detailed procedures that specify which behaviors should be performed and how. In situations where it is unclear exactly which behaviors are effective, such as with a big-ticket sale, managers can use output control strategies. These include setting agreed-upon, measurable goals, such as sales quotas. Alternatively, clan control strategies attempt to control behavior through shared values and beliefs created through selection, training, and socialization.[60]

> **Virtual Wisdom ▼**
>
> *One of the basic tenets of the Cisco culture is trust. If you can't tell what an employee is doing without seeing them physically planted in a chair, then how are you running your business?*
>
> —Bill Finkelstein,
> senior partner at Cisco[61] ▲

All three strategies are useful in helping you to control remote workers. Behavioral control strategies, however, are both inefficient and less effective when workers are not physically present. For this reason, many organizations emphasize outcome control strategies. However, many telecommuting experts suggest focusing on clan strategies.[62] This makes sense because clan strategies work to create internalized change in employees. That is important because we seek to control what we do not trust.[63] If internalized change is created in employees, that is, if they are really "buying into" management goals and values, they will be perceived as more trustworthy by managers, which should reduce fears of losing control over teleworkers.

One way to get employees to "buy in" is to use a management style that encourages employees to participate in decision making, goal setting, and tracking performance. This sort of participatory management also strengthens commitment by sharing information with employees and empowering them.[64]

But encouraging participation may not be enough. Even with the use of clan strategies that engage commitment, smart managers will continue to use outcome strategies, as well. Setting goals must be accompanied by methods of measuring progress toward those goals. In business, progress is measured by deliverables. A deliverable can be any measurable outcome, such as a document outline, sections of a document, prototypes, simulation techniques, and more. Defining frequent deliverables provides a means of monitoring progress and offering feedback early on, if a remote worker is moving in the wrong direction.[65]

Defining and receiving frequent deliverables is another way you can develop trust in your remote employees. Research shows that managers cite job-related characteristics (e.g., competent, highly motivated) as reasons for trusting employees.[66] Frequent deliverables give employees opportunities to demonstrate these characteristics and therefore their trustworthiness.

Just as it is important for managers to trust employees, remote employees must trust managers. Managers contribute greatly to their own perceived trustworthiness by doing what they say they will do. If they effectively manage expectations, are consistent in words and deeds, are fair and stand by their agreements, they will be more trusted.[67] Interestingly, however, employees are likely to offer personal characteristics rather than job-related characteristics when giving reasons for trusting managers.[68] This suggests that to truly strengthen trust, managers must form relationships with employees. Indeed, research shows personal communication and building relationships is a critical part of successful virtual work arrangements[69] and that leaders or managers have an important role in engendering trust between co-workers or team members.[70]

CREATING CONNECTEDNESS FOR REMOTE WORKERS

Building trust and strengthening relationships help create a feeling of being connected to the organization for remote workers who can often feel socially isolated and disconnected. This may help to explain why remote workers are often less committed to their jobs.[71] Creating a sense of organizational identity and connectedness is a challenge for remote workers, but they can accomplish that with some support from their managers. Regular meetings are one way to help create these feelings. Other techniques include the following:

- **Meet face to face.** If possible, an initial face-to-face meeting with remote workers or virtual team members is the best way to form relationships. Periodic face-to-face meetings have also been shown to be useful.[72]
- **Create a visual aid.** It isn't always possible to arrange a face-to-face meeting, but just having a visual image in mind of the people with whom you are communicating can be helpful in developing relationships. Posting pictures of the off-site workers on a web site can be helpful. For virtual teams, a team picture is even better than individual shots. Creating and distributing an object that features a group picture and is likely to be kept around the office in plain view, such as a mousepad or calendar, is a good idea.[73] Groupware is now available that allows you to post pictures and have them pop up when your employee sends a message.
- **Replace the grapevine.** Off-site workers can feel isolated and disconnected from the rest of the organization in part because they are out of touch with informal events in the organization. They miss out on the grapevine, and not just the content of the grapevine, but

Virtual Wisdom ▼

I send a welcome message with everyone's phone number and e-mail address along with a short biography of myself. I send this to the first team member and ask that person to attach his or her bio and send it on, until all members have seen the bios. Create a map of where your virtual team members live, and put the photo on the map so everyone can put a name to a face and location.

—Karen Davidson, President of KDA Software[74] ▲

also the knowledge of what is considered exciting news and what isn't. Having a virtual watercooler or an electronic space where casual conversation may occur is a good way to encourage informal interaction. Distributing a newsletter and written, journalistic-style accounts of what transpired during meetings can be helpful. Mention of remote workers in these documents, such as an acknowledgment of those who are not present at the start of meetings or highlighting stellar performance, is a good strategy for creating feelings of involvement for remote workers and a recognition of their involvement by co-workers located in the traditional office.[75]

- **Encourage interaction.** Computer-mediated conversations between off-site workers can often include only work-related exchanges. But rich, complex conversation about a variety of topics is important in forming relationships. Distributed workers may need encouragement to expand conversation beyond work-related topics. Managers and team leaders, who themselves use humor and personal communication, play a pivotal role in facilitating socialization for remote workers and are perceived as more caring and effective.[76] Lisa Kimball and Amy Eunice recommend other techniques for encouraging rich conversations, including holding teleconferences where no administrative issues are allowed, assigning various employees the responsibility for facilitating a discussion on nonroutine topics, and inviting guests from outside the organization to participate in meetings that engage the workers.[77] Interaction among remote workers can lead to strengthened relationships and even friendships. Having strong relationships and friendships at work leads to higher productivity and worker retention levels. In addition, friends are able to challenge one another's ideas in a way that people who don't know each other well may feel is impolite.[78]

The techniques, used in combination with others we discussed earlier, such as scheduling regular meetings and using a variety of communication technologies, can help foster a sense of connectedness in remote workers. It's also helpful to use common sense and consideration for remote workers who are not able to walk down the hall to seek information. One telecommuter tells the story of a planned teleconference she was to participate in. She sat and waited for a conference call that never came. Later she discovered that the meeting had actually taken place, but that the other participants decided she didn't need to be included. Because that decision was handled so badly, the telecommuter had no idea if it was an innocent decision (meaning she really didn't need to participate, and the others didn't want to waste her time) or if it was devious (meaning they didn't really want her input to begin with). She isn't able to walk down the hall and ask, "What's going on?" and she has no way of inferring intentions from reading the body language of those that decided to exclude her. Obviously, if she felt connected before this incident, she certainly would not have felt connected after it.[79]

FACILITATING, MOTIVATING, AND EVALUATING PERFORMANCE

Clear communication from management about organizational and team goals is essential for successful performance of remote workers. Managers facilitate performance by providing the resources necessary to complete a job and by eliminating barriers to effectiveness. In a virtual work situation, both of these largely involve providing good equipment and technical support. We've already talked about the importance of providing better equipment. It is unreasonable to expect a remote worker to perform well without the proper tools for the job. Barriers to effectiveness can include outdated equipment, lack of training, delays in receiving information, inaccessibility of information, or work processes that unnecessarily impede performance. For example, a form that needs to be signed by several people—for no reason other than tradition—could be easy enough for an on-site worker to accept but could be an awesome burden for an off-site worker. The workers themselves are usually a great source of information about barriers that inhibit their performance.[80]

Motivating employees to perform is a complex process that involves so many factors. But one fairly simple way to motivate them is to reward performance. This is only *fairly* simple because what is rewarding to one employee may not be rewarding to another. To properly reward behavior, managers must understand what is important to each employee. Then, reward structures can be tailored to each employee's values. Distribution of rewards, however, must be perceived as fair and thus cannot be too individualized.[81]

Virtual Wisdom ▼

Organizations are beginning to give workers more flexibility, autonomy, and accountability to design rewards. Workers get to decide whether they want two, all, or none of the items—such as skill-based pay, team incentives, team recognition, and goal-sharing—that are part of the reward menu.
—Peter V. LeBlanc, National Practice Leader, Sibson & Co.[82] ▲

Many times remote workers don't get the informal praise that on-site workers receive. This isn't a conscious neglect from managers; it's simply the result of not having opportunities for casual communication at the drinking fountain or the chance to lean in an office doorway and offer a word of praise. Giving recognition is a way to motivate and energize employees, so consciously planning to give praise to remote workers is a valuable practice. Employees who work in the office may have little knowledge of the accomplishments of off-site workers because of the lack of visibility and social presence. Creating the social presence of remote workers by including their pictures on the office bulletin board and their names on the mailboxes are just a couple of good ways to encourage their remembrance by on-site workers.[83]

Modern performance evaluation techniques typically offer a set of skills and characteristics that are deemed necessary to complete a job on which employees are to be evaluated. Managers then observe these skills and characteristics in the workplace, and these observations become the basis of performance evaluations. Measures can be subjective, such as a rating by a manager, or they can be objective, such as a count of the number of units produced. Most traditional measures of performance are subjective measures (e.g., ratings of traits) of work processes. This approach has some obvious limitations for managers in the virtual workplace. To evaluate teleworkers, managers measure the quality of the outcomes and the quantity of the work and move away from focusing on how the work is done.[84]

Elizabeth Allen, Vice President of Corporate Communications at Dell, enlists the aid of local managers when evaluating members of her global virtual team. Local managers can help her get a sense of the team member's attitude at work, which she can't observe. Even though

Dell is an "e-mail culture," Allen prefers face-to-face contact for providing performance evaluation.[85]

TRAINING AND MENTORING

Numerous e-learning tools are available for companies to use in training employees—on- and off-site. Organizations are discovering that Internet-based training methods are easy to access and are cost efficient. E-learning allows training programs to be customized so that they are learner-centered. Employees can engage in continuous learning from anywhere, anytime. These tools are very effective in developing employees' technical skills.[86] Martha Haywood recommends that remote workers be trained in communication skills, communication technology skills, leadership skills, and management skills, as well.[87] Virtual team members can also benefit from training in team processes.

Mentors provide remote workers with a valuable opportunity for employee development. They can provide feedback, access to networks, and emotional support. Mentoring remote workers, of course, is challenging because the employees' behavior isn't observable; mentors may have a difficult time discerning how employees are really doing and what their needs are. Mentors may also feel that employees are less likely to communicate via mediated communication when things may be bothering them. Periodic face-to-face meetings or phone conversations can help solve this problem.[88]

Managers of virtual teams complain that it's difficult to transfer the informal interactive learning that occurs in the office to remote workers. This type of learning isn't formally scheduled and offers employees a sort of *in the moment* career development. Managers can compensate for some of this loss by creating virtual watercoolers, assigning mentors, and having informal electronic chats.[89]

Training managers is just as important as training employees. The management of telecommuters cannot be approached in the same way as traditional management. Preparing through training can mean the difference between the success and failure of a telecommuting program.[90] Traditional methods of managing by walking around are irrelevant in a virtual organization. Managers must shift from a task-and-time orientation to a results focus, and this transition can be difficult. But virtual management training can help, and offer unexpected benefits. Many managers find that they are better managers for on-site employees after spending time training to manage off-site employees. So, management training, which many managers may resist, could become more agreeable if it is in the form of something new, such as training to manage an off-site team.[91]

Virtual Wisdom ▼

Teamwork today is trickier than ever.
—Laurie Putnam, Manager of Information
Design, Aspect Communications[92] ▲

EFFECTIVE VIRTUAL TEAMS

Not only do virtual team members need all the skills and training that other remote workers need, they also need team skills. All the problems that occur with co-located teams also occur in distributed teams, but they are often much greater or more serious. Virtual teams that are not properly managed can take longer to produce results than traditional teams and are often less accurate in their information sharing.[93] For that reason alone, taking the following steps to build and manage distributed teams becomes all the more important.

- **Instill a sense of purpose.** Having a mission, clear goals, and specific tasks will contribute to a strong sense of purpose. An understanding of the team's mission and goals will serve to guide team members in ambiguous situations, which is particularly important since they can't ask questions over lunch. Consensus on mission and goals is particularly important. Consensus does not mean that each person on the team thinks that the solution the team selects is the best one. But it does mean that each member of the team can live with the solution.[94]

Virtual Wisdom ▼

There needs to be alignment around purpose, dealing with conflict, sharing leadership. These are really significant problems with teams, and we've barely figured out how to solve them face to face, never mind when people are spread out around the globe.

—Jessica Lipnack, CEO, VirtualTeams.com[95] ▲

- **Limit the size of the team.** Even though a virtual team does not have to deal with the issues of how big the circumference of the table needs to be to fit all the team members around it, teams become inefficient and cease to function as teams if there are too many people included. A team should have as many members as it takes to do the job. Too many members on a team can promote social loafing, which occurs when one or more team members lets the other members carry their weight. It also promotes the formation of subgroups, a questioning of purpose ("Why am I part of this group when there is nothing for me to do?"), and a disconnect from the team whose members are underutilized.[96]
- **Establish leadership.** If a leader has not been pre-assigned, team members should choose their leader in the early stages of team formation. A leader at this stage is useful because, during initial team formation, people are unsure of how to behave and want to be polite and present their best social image. Team members will typically not ask questions or raise issues at this stage. Team members may not be want to be seen as controlling or pushy by taking a directive role at the start. The team will move through this initial stage and become more productive more quickly if someone is assigned the leadership role, if only temporarily. Eventually, leadership should be shared, and the role of leader should be assigned based on the fit of the person with the task at hand.[97]
- **Be clear about members' roles.** In a virtual team, members may or may not have access to each other's work. If the members cannot see each other's work, then being very clear about roles and responsibilities leads to better coordination of effort and decreases the likelihood of members duplicating each other's work.[98]
- **Provide training.** Teams that have at least one member that has had some training in the development of teams have better outcomes than teams that do not. This is true for virtual

teams as well as co-located teams. For virtual teams, additional training in communication technologies and groupware is also helpful, along with instruction on how to use this software to provide anonymous feedback. Training on organizational culture may be valuable, as well. If teams are created across organizations, the team may generate a team culture.[99]

- **Spell out everything.** We often believe that a single word means the same thing to everyone who hears it. This is, of course, not true and is especially important to remember when no verbal or other contextual cues are available to help a receiver correctly interpret a message. For example, the word *quality* may mean different things to different people, especially if they are from different organizations or even different functional areas.[100]

- **Have teams develop a plan for process.** Team members should share their expectations and decide how they will make decisions, how they will communicate, and how they will resolve conflict. It can be useful to have your teams draft a contract about these matters that each team member agrees to follow. Managers can then monitor the team's progress based on the processes and outcomes they've agreed upon.[101]

- **Consider individual goals.** The success of any team depends on the members' willingness to put team goals ahead of their individual agendas—at least sometimes. However, it's important to recognize that over the long run individual team members must be able to meet their individual goals, while meeting team goals at the same time. You should address this issue openly so that it becomes a non-issue and team members can help each other achieve individual goals. Team members must be accountable for both the success of the team and the completion of their individual assignments.[102]

- **Communicate outside the team.** Teams that don't communicate with others in an organization can be perceived negatively. Part of successfully managing a virtual team means providing them with communication plans for contacting people elsewhere in the organization and keeping them informed of important developments affecting the organization. A benefit that teams can provide is sharing their best practices with the rest of the organization. When you discover things that work well, share these discoveries with other teams in the organization.[103]

- **Welcome new members personally.** When a new team member is added to an existing team, the team leader should welcome the new member face to face or by telephone, carefully discussing the team's mission and goals and the team's norms. The roles and expectations of the new member should be thoroughly explained. A new member will benefit from having a partner assigned from the team who can facilitate the orientation and review how to use the hardware and software the group uses. Virtual team members thus have an opportunity that traditional team members don't have, in that they can review

Virtual Wisdom ▼

We have an obligation to meet the needs of business and be strategic partners. As we attract, retain, and develop the best talent, we have to assess employees on a continuing basis for flexibility and adaptability to work in a virtual environment—that is the 21st century.

—Joy Gaetana, Senior Vice President of Corporate Human Resources, USFilter[104] ▲

records—if they they've been kept—of previous meetings and interactions among existing team members.[105]

The most critical thing to remember about virtual teams is that they are, first and foremost, teams, and all the rules for creating effective teams apply to them as well.

WORKING EFFECTIVELY OFF-SITE

Working off-site and liking it require a combination of characteristics. According to Silvia Orian from Xerox, you must be self-disciplined. She is up, dressed, and on her computer every morning and sees clients every afternoon. You also need to be independent and not require a lot

---

Virtual Wisdom ▼

Just between you and me, I can strongly recommend telecommuting. I'm writing this article from my home-office 100 kilometres away from the nearest capital city. My colleagues can't complain about my bad taste in music—because they can't hear it. I've got a cozy log-fire burning by my side, and I may as well be wearing my PJs since it's 3 a.m. But as long as this story is e-mailed by 9 a.m., my employer will never know about my all-night stint, wink-wink, nudge-nudge!
—Elizabeth Walton, business writer[106] ▲

---

of social interaction with co-workers. Some telecommuters experience loneliness and a sense of isolation. Orian recommends that telecommuters stay in touch with other professionals through organizations like Business Network International and the local Chamber of Commerce. Orian also recommends that telecommuters know their IT. Learn how to troubleshoot on your own so you don't experience long periods of downtime waiting for help.[107]

While IT problems have been Orian's greatest frustration working from home, other telecommuters have experienced a variety of challenges. Stephen, a programmer in Silicon Valley, convinced his boss to let him work from home a few days a week. Very soon into the arrangement, Stephen became aware of some problems that were poisoning his productivity. Here are some of his challenges and eventual solutions:[108]

- **Too many distractions.** Some people are distracted by the complete silence of home, so they turn on the TV, and then they are distracted by the show. Some people are able to reach a workable compromise by listening to music, played softly.
- **Unwanted visitors.** Friends and family, when they find out you're home, might want to stop by for a chat. You have to lay down the law. Set some rules about interruptions.
- **Irresistible snacks.** For this, there is no solution. You could avoid buying your favorite snacks, but even saltines become irresistible when you're bored and want a reason to take a break.
- **Space issues.** Sometimes a space that seemed large before the office equipment is moved in can seem small afterward. Use space creatively.
- **Distracting pets.** If you can't tune out the dogs chasing each other around the room while you're working, you may have to put them in another room.

Once you have worked out the issues at home, you still want be as effective as possible at the office, even though you're not there. Stay "visible" by communicating regularly, even if this means you must take the initiative. Keep your boss updated so he or she doesn't have to wonder what you're doing. Work on your communication skills if you recognize any skill gaps. And remember, your communication skills include not only your ability to write a concise e-mail or leave a clear voice mail, but also to choose the best medium and to anticipate how your audience will interpret your message without many nonverbal cues. When you are an off-site employee, your verbal messages represent you.[109]

Whether you are the manager of off-site workers, a member of a virtual team, or a telecommuter a few days a week, the most important thing you can do as part of a virtual work environment is to communicate effectively. Electronic communication networks are the circulatory system of the virtual organization, and the organization can't survive without a regular flow of effective messages. Dependable communication, carefully composed messages, and appropriate media choices can help those in virtual work environments achieve communication and organizational objectives. Only with these effective communication practices are building relationships and trust in virtual organizations possible.

Virtual work environments allow the best minds, the most creative thinkers, and the most desired experts to work together, while apart. Through the use of advanced communication technologies, virtual organizations can gain competitive advantage by transcending the limits of time and distance. But the technology alone is not enough. A successful virtual work environment develops after careful planning and preparation. Remote workers and those that manage them must learn how to use communication and communication technology effectively and strategically to accomplish organizational objectives. While this is important in all organizations, it is vital in virtual work environments, because communication literally holds virtual organizations and those that work in them together.

DISCUSSION QUESTIONS

1. One of your managers has just been assigned to manage a group of telecommuters. He comes to you and expresses his concern about how he will know whether they are putting in 40 hours per week. What do you say to him?

2. Think of parts of your most recent job that could be performed from a remote location. In what ways would you measure the quality and quantity of performance of that job?

3. What ways can you think of to boost the visibility and connectedness of remote workers? What can you do to make them feel more a part of the organization?

4. What strategies would you suggest for managing conflict between virtual team members who are not able to meet face to face?

ENDNOTES

1. Wayne F. Cascio, "Managing a Virtual Workplace," *Academy of Management Executive,* vol. 14 (2000): 82–90.
2. Susan J. Wells, "Making Telecommuting Work," *HR Magazine,* vol. 46 (2001): 34–46.

3. Kemba J. Dunham, "Telecommuters' Lament," *Wall Street Journal,* 31 October 2000, B1, B18.
4. Catherine Siskos, "Out of Sight . . . ," *Kiplinger's Personal Finance Magazine,* vol. 56 (January 2002): 26.
5. Cascio.
6. Warren S. Hersch, Jennifer O'Herron, and Jackie Taylor, "Primetime for Telecommuting Anytime, Anywhere," *Call Center Magazine,* vol. 14 (October 2001): 34–43.
7. Renuka Rayasam and Monica Stevenson, "Telecommuting Is a Flop So Fix It Already," *Ziff Davis Smart Business,* vol. 14 (August 2000): 64–67.
8. Carl E. Van Horn and Duke Storen, "Telework: Coming of Age? Evaluating the Potential Benefits of Telework," *Telework and the New Workplace of the 21st Century,* 2000. Available: *http://www.dol.gov/asp/telework/p1_1.htm.*
9. Cascio.
10. Claire R. McInerney, "Working in the Virtual Office: Providing Information and Knowledge to Remote Workers," *Library and Information Science Research,* vol. 21 (1999): 69–89.
11. Rebecca R. Hastings, "Sample Telecommuting Policy," Society for Human Resource Management, 1999. Available: *http://www.shrm.org.*
12. Wells.
13. Ibid.
14. Cascio.
15. George G. Baffour and Charles L. Betsey, "Human Resources Management and Development in the Telework Environment," *Telework and the New Workplace of the 21st Century,* 2000. Available: *http://www.dol.gov/asp/telework/p2_2.htm.*
16. Cascio.
17. Charlotte Garvey, "Teleworking HR," *HR Magazine,* vol. 46 (2001): 56–61.
18. Ibid.
19. Hastings.
20. Wells.
21. Silvia Orian, personal communication with author, June 2002.
22. Cascio.
23. Katie Hafner, "Working at Home Today?" *New York Times,* 2 November 2001, D1, D8.
24. Rayasam and Stevenson, 64–67.
25. Ibid.
26. Wells.
27. Baffour and Betsey.
28. Garvey.
29. Wells.
30. Martha Haywood, *Managing Virtual Teams: Practical Techniques for High-Technology Project Managers* (Boston: Artech House, 1998), 65–68.
31. Haywood.
32. Caron Schwartz Ellis, "Telecommuting Quickly Becoming Benefit for Employer, Employee," *Boulder County Business Report,* 1995. Available: *http://bcn.boulder.co.us/business/BCBR/1995/sep/commute2.html.*
33. William R. Pape, "Hire Power," *Inc.,* 2002. Available: *http://www.inc.com.*
34. Deborah L. Duarte and Nancy T. Snyder, *Mastering Virtual Teams* (San Francisco: Jossey-Bass, 1999).
35. Ibid.
36. Hersch, O'Herron, and Taylor, 34–43.
37. Renuka and Stevenson.
38. Marianne K. McGee, "The Home Front," *Information Week,* 22 October 2001, 55–58.
39. Pape.

40. Michael J. Demaria, "Telecommuting: Keeping Data Safe and Secure," *Network Computing,* vol. 12 (2001): 87–91.
41. Ibid.
42. Haywood.
43. Michael Verespej, "The Compelling Case for Telework," *Industry Week,* vol. 250 (September 2001): 23.
44. Michael Verespej, "The Old Workforce Won't Work," *IndustryWeek.com,* September 1998. Available: *http://www.industryweek.com.*
45. Hastings.
46. Hersch, O'Herron, and Taylor.
47. Orian.
48. Haywood.
49. Sandi Mann, Richard Varey, and Wendy Button, "An Exploration of the Emotional Impact of Teleworking via Computer-Mediated Communication," *Journal of Managerial Psychology,* vol. 15 (2000): 668–690.
50. Haywood.
51. Wells.
52. Sheila M. Bunting, Cynthia K. Russell, and David M. Gregory, "Use of Electronic Mail for Concept Synthesis: An International Collaborative Project," *Qualitative Health Research,* vol. 8 (1998): 128–136.
53. Linda Duxbury and Derrick Neufeld, "An Empirical Evaluation of the Impacts of Telecommuting on Intra-Organizational Communication," *Journal of Engineering and Technology Management,* vol. 16 (1999): 1–28.
54. Paul Falcone, "Reinventing the Staff Meeting," *HR Magazine,* August 2000, 143–146.
55. Elizabeth Kelley, "Keys to Effective Virtual Global Teams," *Academy of Management Executive,* vol. 15 (2001): 132–134.
56. Lisa Kimball and Amy Eunice, "The Virtual Team: Strategies to Optimize Performance," *Health Forum Journal,* vol. 42 (1999): 58–63.
57. "Communication in Virtual Teams," *Strategic Communication Management,* vol. 5 (2001): 3.
58. Haywood.
59. Timothy Kayworth and Dorothy Leidner, "The Global Virtual Manager: A Prescription for Success," *European Management Journal,* vol. 18 (2000): 183–189.
60. Nancy B. Kurland and Cecily D. Cooper, "Manager Control and Employee Isolation in Telecommuting Environments," *The Journal of High Technology Management Research,* vol. 13 (2002): 107–126.
61. Hafner.
62. Kurland and Cooper.
63. Carole V. Wells and David Kipnis, "Trust, Dependency, and Control in the Contemporary Organization," *Journal of Business and Psychology,* vol. 15 (2001): 593–603.
64. Haywood.
65. Ibid.
66. Wells and Kipnis.
67. "Creating Successful Virtual Teams," *Harvard Management Communication Letter,* vol. 3 (December 2000): 10–12.
68. Wells and Kipnis.
69. Jeremy S. Lurey and Mahesh S. Raisinghani, "An Empirical Study of Best Practices in Virtual Teams," *Information and Management,* vol. 38 (2001): 523–544.
70. Kayworth and Leidner.
71. Hersch, O'Herron, and Taylor.
72. Lurey and Raisinghani.

73. Kimball and Eunice.
74. Carla Joinson, "Managing Virtual Teams, "*HR Magazine,* vol. 47 (2002): 69–73.
75. Kimball and Eunice.
76. Kayworth and Leidner.
77. Kimball and Eunice.
78. Susan Ellingwood, "The Collective Advantage," *Gallup Management Journal,* vol. 1 (2001): 6–8.
79. Dunham.
80. Cascio.
81. Ibid.
82. Verespej, 1998.
83. Kimball and Eunice.
84. Baffour and Betsey.
85. Elizabeth Allen, personal communication with author, June 24, 2002.
86. "Trends: Training for the Future," November 15, 2000. Available: *http://www.businesscisco.com.*
87. Haywood.
88. Kurland and Cooper.
89. Ibid.
90. Wells.
91. Cascio.
92. Laurie Putnam, "Distance Teamwork," *Online,* vol. 25 (2001): 54–58.
93. Kenneth A. Graetz, Edward S. Boyle, Charles E. Kimball, Pamela Thompson, and Julie L. Garlouch, "Information Sharing in Face-to-Face Teleconferencing and Electronic Chat," *Small Group Research,* vol. 29 (1998): 714–744.
94. Charles Wardell, "The Art of Managing Virtual Teams: Eight Key Lessons," *Harvard Management Update,* vol. 3 (1998): 4–6.
95. Julekha Dash, "Think of People When Planning Virtual Teams," *Computerworld,* vol. 35 (February 2001): 34.
96. Sharon A. Wheelan, *Creating Effective Teams* (Thousand Oaks, CA: Sage, 1999).
97. Ibid.
98. Ibid.
99. Carla Joinson, "Managing Virtual Teams," *HR Magazine* (June 2002): 69–73.
100. Wheelan.
101. Wardell.
102. Wheelan.
103. Ibid.
104. Duarte and Snyder.
105. Wardell.
106. Elizabeth Walton, "Virtual Office," Available: *http://www.pnc.com.au/~lizzi/Virtual_office.html.*
107. Orian.
108. Joanne Eglash, "Virtual Office," 2000. Available: *http://careerlink.devx.com/articles/je062600.asp.*
109. Putnam.

APPENDIX

A SELECT BIBLIOGRAPHY

Adams, T., and N. Clark. *The Internet: Effective Online Communication.* Fort Worth, TX: Harcourt College Publishers, 2001.

Ahuja, Manju K., and Kathleen M. Carley. "Network Structure in Virtual Organizations." *Journal of Computer-Mediated Communication,* vol. 3 (1998). Available: *http://www.ascusc.org/jcmc/vol3/issue4/ahuja.html.*

Anderson, Ron. "Sharing Is Daring." *Network Computing,* vol. 13 (February 2002): 47–61.

Baffour, George G., and Charles L. Betsey. "Human Resources Management and Development in the Telework Environment." *Telework and the New Workplace of the 21st Century,* 2000. Available: *http://www.dol.gov/asp/telework/p2_2.htm.*

Barkhi, Reze, Varghese S. Jacob, and Hasan Pirkul. "An Experimental Analysis of Face-to-Face Versus Computer-Mediated Communication Channels." *Group Decision and Negotiation,* vol. 8 (1999): 325–347.

Bavelas, Janet Beavin, and Nicole Chovil. "Visible Acts of Meaning." *Journal of Language and Social Psychology,* vol. 19 (2000): 163–195.

Bean, Sarah. "How to . . . Bring Rita & Rob Together." *Enterprise* (April/May 2002): 21–23.

Beebe, Steven, Susan Beebe, and Mark Redmond. *Interpersonal Communication,* 2nd ed. Boston: Allyn and Bacon, 1999, 11–16.

Bunting, Sheila M., Cynthia K. Russell, and David M. Gregory. "Use of Electronic Mail for Concept Synthesis: An International Collaborative Project." *Qualitative Health Research,* vol. 8 (1998): 128–136.

Byfield, Mike. "Personal Video Calling Is Ready for Prime Time." *Report/Newsmagazine* (Alberta Edition), vol. 29 (2002): 40–41.

Captain, Timothy. "Remote Access." *Laptop* (July 2002): 22–30.

Cascio, Wayne. "Managing a Virtual Workplace." *Academy of Management Executive,* vol. 13, no. 3 (2000): 81–90.

Cooper, W. W., and M. L. Muench. "Virtual Organizations: Practice and Literature." *Journal of Organizational Computing & Electronic Commerce,* vol. 10 (2000): 189–209.

Daft, R. L., R. H. Lengel, and L. Trevino. "Message Equivocality, Media Selection, Manager Performance." *MIS Quarterly,* vol. 11 (1987): 355–366.

Demaria, Michael J. "Telecommuting: Keeping Data Safe and Secure." *Network Computing,* vol. 12 (2001): 87–91.

Dennis, Alan R., and Susan T. Kinney. "Testing Media Richness Theory in the New Media: The Effects of Cues, Feedback, and Task Equivocality." *Information Systems Research,* vol. 9 (September 1998): 256–275.

De Tienne, Kristen Bell. *Guide to Electronic Communication.* Upper Saddle River, NJ: Prentice Hall, 2002.

Duarte, Deborah L., and Nancy T. Snyder. *Mastering Virtual Teams.* San Francisco: Jossey-Bass, 1999.

Dunham, Kemba J. "Telecommuter's Lament." *Wall Street Journal,* 31 October 2000, B1, B18.

Duxbury, Linda, and Derrick Neufeld. "An Empirical Evaluation of the Impacts of Telecommuting on Intra-Organizational Communication." *Journal of Engineering and Technology Management,* vol. 16 (1999): 1–28.

Ekman, Paul. "Facial Expression and Emotion," *American Psychologist,* vol. 48 (1993): 384–392.

Ekman, Paul, and Maureen O'Sullivan. "Who Can Catch a Liar." *American Psychologist,* vol. 46 (1991): 913–920.

Elashmawi, Farid. "Overcoming Multicultural Clashes in Global Joint Ventures," Available: *http://www.iftdo.org/articles.htm.*

Ellingwood, Susan. "The Collective Advantage." *Gallup Management Journal,* vol. 1 (2001): 6–8.

Ellis, Caron Schwartz. "Telecommuting Quickly Becoming Benefit for Employer, Employee." *Boulder County Business Report,* 1995. Available: *http://bcn.boulder.co.us/business/BCBR/1995/sep/commute2.html.*

Eom, S., and C. Lee. "Virtual Teams: An Information Age Opportunity for Mobilizing Hidden Manpower." *S.A.M. Advanced Management Journal,* vol. 64 (1999): 12–17.

Falcone, Paul. "Reinventing the Staff Meeting." *HR Magazine* (August 2000): 143–146.

Fane-Saunders, Terence. "Public Relations in Asia: The International Dimension." 22 February 2002. Available: *http://www.issa.com/features/intl/2002feb22.html.*

Fatt, James P. T. "It's Not What You Say, It's How You Say It." *Communication World,* vol. 16 (June/July 1999): 37–41.

Feingold, A. "Good-looking People Are Not What We Think." *Psychological Bulletin,* vol. 111 (1992): 304–341.

Fulk, Janet, and Lori Collins-Jarvis. "Wired Meetings," in *The New Handbook of Organizational Communication,* edited by Fredric M. Jablin and Linda Putnam. Thousand Oaks, CA: Sage, 2001, 624–663.

Garvey, Charlotte. "Teleworking HR." *HR Magazine,* vol. 46 (2001): 56–61.

Gibbs, W. Wayt. "The Network in Every Room." *Scientific American,* vol. 286, no. 2 (February 2002): 38–53.

Glater, Jonathan. "Telecommuting's Big Experiment: Federal Employees Are Urged to Work Outside of the Office." *New York Times,* 9 May 2001, C1.

Gould, David. "Virtual Organization," 2000. Available: *http://www.seanet.com/~daveg.*

Grabbe, Nick. "Telecommuting Works for Many," 7 February 2000. Available: *http://www.gazettenet.com/biz2000/02072000/21636.htm.*

Graetz, Kenneth A., Edward S. Boyle, Charles E. Kimball, Pamela Thompson, and Julie L. Garlouch. "Information Sharing in Face-to-Face Teleconferencing and Electronic Chat," *Small Group Research,* vol. 29 (1998): 714–744.

Grundy, John. "Trust in Virtual Teams." *Harvard Business Review,* vol. 76 (1998): 180.

Guiliano, Peter. "Seven Benefits of Eye Contact." *Successful Meetings,* vol. 48 (1999): 104.

Gunn, John. "The Virtual Corporation," December 2001. Available: *http://bcn.boulder.co.us/business/BCBR/december/virtual.dec.html.*

Hafner, Katie. "Company Sites Are Holding Offices Together." *New York Times,* 26 September 2001, D1, D8.

Hafner, Katie. "Working at Home Today?" *New York Times,* 2 November 2001, D1.

Hancock, Jeffrey T., and Philip J. Dunham. "Impression Formation in Computer-Mediated Communication Revisited." *Communication Research,* vol. 28 (2001): 325–347.

Hastings, Rebecca R. "Sample Telecommuting Policy." Society for Human Resource Management, 1999. Available: *http://www.shrm.org.*

Hawkins, Dana. "Lawsuits Spur Rise in Employee Monitoring." *U.S. News & World Report,*" 13 August 2001, 53.

Haywood, Martha. *Managing Virtual Teams: Practical Techniques for High-Technology Project Managers.* Boston: Artech House, 1998.

Hersch, Warren S., Jennifer O'Herron, and Jackie Taylor. "Primetime for Telecommuting Anytime, Anywhere." *Call Center Magazine,* vol. 14 (October 2001): 34–43.

Jarvenpaa, S., and D. Leidner. "Communication and Trust in Global Virtual Teams." *Organization Science: A Journal of the Institute of Management Sciences,* vol. 10 (1999): 791–846.

Johnson, Dave. "Smartphone Platform Wars." *Laptop* (July 2002): 32–46.

Joinson, Carla. "Managing Virtual Teams." *HR Magazine* (June 2002): 69–73.

Kasper-Fuehrer, Eva C., and Neal M. Ashkanasy. "Communicating Trustworthiness and Building Trust in Interorganizational Virtual Teams." *Journal of Management,* vol. 27 (2001): 238.

Kayworth, Timothy, and Dorothy Leidner. "The Global Virtual Manager: A Prescription for Success." *European Management Journal,* vol. 18 (2000): 183–189.

Kelleher, Joanne. "E-meetings Redefine Productivity." *Fortune,* 5 February 2001, S2–S12.

Kelley, Elizabeth. "Keys to Effective Virtual Global Teams." *Academy of Management Executive,* vol. 15 (2001): 132–134.

Kimball, Lisa, and Amy Eunice. "The Virtual Team: Strategies to Optimize Performance." *Health Forum Journal,* vol. 42 (1999): 58–63.

Kraus, R. M., and W. Weinheimer. "Concurrent Feedback, Confirmation, and the Encoding of Referents in Verbal Communication." *Journal of Personality and Social Psychology,* vol. 4 (1966): 343–346.

Kurland, Nancy B., and Cecily D. Cooper. "Manager Control and Employee Isolation in Telecommuting Environments." *The Journal of High Technology Management Research,* vol. 13 (2002): 107–126.

LaPlante, Alice. "90s Style Brainstorming." *Forbes,* vol. 152 (October 1993): 44–49.

Lovelace, Glenn. "The Nuts and Bolts of Telework: Growth in Telework." *Telework and the New Workplace of the 21st Century,* 2000. Available: *http://www.dol.gov/asp/telework/p1_2.htm.*

Lurey, Jeremy S., and Mahesh S. Raisinghani. "An Empirical Study of Best Practices in Virtual Teams." *Information and Management,* vol. 38 (2001): 523–544.

Mann, Sandi, Richard Varey, and Wendy Button. "An Exploration of the Emotional Impact of Teleworking via Computer-Mediated Communication." *Journal of Managerial Psychology,* vol. 15 (2000): 668–690.

McGee, Marianne K. "The Home Front." *Information Week,* 22 October 2001, 55–58.

McInerney, Claire R. "Working in the Virtual Office: Providing Information and Knowledge to Remote Workers." *Library and Information Science Research,* vol. 21 (1999): 69–89.

McLuhan, Marshall. *Understanding Media: The Extensions of Man.* New York: The New American Library, 1964.

Mead, Dana G. "Retooling for the Cyber Age." *CEO Series,* 42, September 2000. Available: *http://csab.wustl.edu/csab/.*

Mehrabian, A. "Verbal and Nonverbal Interaction of Strangers in a Waiting Situation." *Journal of Experimental Research in Personality,* vol. 5 (1971): 127–138.

Mennecke, Brian E., Joseph S. Valacich, and Bradley C. Wheeler. "The Effects of Media and Task on User Performance: A Test of the Task-Media Fit Hypothesis." *Group Decision and Negotiation,* vol. 9 (2000): 507–529.

Mirchandani, Kiran. "Legitimizing Work: Telework and the Gendered Reification of the Work/Nonwork Dichotomy." *Canadian Review of Sociology and Anthropology,* vol. 36 (1999): 87–107.

O'Rourke, IV, James S. *Management Communication: A Case-Analysis Approach* (pp. 21–22). Upper Saddle River, NJ: Prentice Hall, 2001.

Overholt, Alison. "Virtually There?" *Fast Company* (March 2002): 108.

Palmer, Jonathan W., and Cheri Speier. "A Typology of Virtual Organizations: An Empirical Study," 1997. The University of Oklahoma. Available: *http://hsb.baylor.edu/ramsower/ais.ac.97/papers/palm_spe.htm.*

Pape, William R. "Hire Power." *Inc.,* 2002. Available: *http://www.inc.com.*

Perry, John. "Palm Power in the Workplace." *American Salesman,* vol. 46 (2001): 22, 5p.

Postmes, Tom, Russell Spears, and Martin Lea. "The Formation of Group Norms in Computer-Mediated Communication." *Human Communication Research,* vol. 26 (2000): 341–371.

Purdy, Jill M. "The Impact of Communication Media on Negotiation Outcomes." *International Journal of Conflict Management,* vol. 11 (2000): 162–184.

Putnam, Laurie. "Distance Teamwork." *Online,* vol. 25 (2001): 54–58.

Rayasam, Renuka, and Monica Stevenson. "Telecommuting Is a Flop So Fix It Already." *Ziff Davis Smart Business,* vol. 14 (August 2000): 64–67.

Riley, Patricia, Anu Mandavilli, and Rebecca Heino. "Observing the Impact of Communication and Information Technology on Net-Work." *Telework and the New Workplace of the 21st Century,* 2000. Available: *http://www.dol.gov/asp/telework/p2_3.htm.*

Rouseau, D. M., S. B. Sitkin, R. S. Burt, and C. Camerer. "Not So Different After All: A Cross-Discipline View of Trust." *Academy of Management Review,* vol. 23 (1998): 393–404.

Russo, J., and P. Schoemaker. *Decision Traps.* New York: Doubleday, 1989.

Sacash, David. "E-Mail and the CEO." *Industry Week,* vol. 250 (11 June 2002): 29–32.

Shani, A. B. Rami, and James B. Lau. *Behavior in Organizations.* Boston: McGraw-Hill Higher Education, 2000, 303.

Short, J., E. Williams, and B. Christie. *The Social Psychology of Telecommunications.* London: Wiley, 1976.

Siskos, Catherine. "Out of Sight . . . ." *Kiplinger's Personal Finance Magazine,* vol. 56 (January 2002): 26.

Spears, Russell, and Lea Martin. "Panacea or Panopticon?" *Communication Research,* vol. 21 (1994): 427–460.

Strom, David. *Web-Based Discussion Forum Software,* 2002. Available: *http://www.strom.com/places/wc.html.*

Sussman, Stephanie, and Lee Sproull. "Straight Talk: Delivering Bad News through Electronic Communication." *Information Systems Research,* vol. 10 (June 1999): 150–167.

Thio, Alex. *Sociology,* 5th ed. New York: Addison Wesley Longman, 1998.

Townsend, Anthony M., Samuel M. DeMarie, and Anthony R. Hendrickson. "Virtual Teams: Technology and the Workplace of the Future." *Academy of Management Executive,* vol. 12, no. 3 (1998): 17–29.

Van Horn, Carl E., and Duke Storen. "Telework: Coming of Age? Evaluating the Potential Benefits of Telework." *Telework and the New Workplace of the 21st Century,* 2000. Available: *http://www.dol.gov/asp/telework/p1_1.htm.*

Verespej, Michael A. "The Compelling Case for Telework." *Industry Week,* vol. 250 (September 2001): 23.

Verespej, Michael A. "The Old Workforce Won't Work." *IndustryWeek.com,* September 1998. Available: *http://www.industryweek.com.*

Vinas, Tonya. "Meetings Makeover." *eBusiness* (February 2002): 29–35.

Voigt, Kevin. "Some Enterprising Employees Exchange the Cubicle Life for Pools in Exotic Places." *Wall Street Journal,* 31 January 2001, B1.

Walther, Joseph B., Celeste L. Slovacek, and Lisa C. Tidwell. "Is a Picture Worth a Thousand Words?" *Communication Research,* vol. 28 (2001): 105–134.

Walton, Elizabeth. "Virtual Office," Available: *http://www.pnc.com.au/~lizzi/Virtual_office.html.*

Wardell, Charles. "The Art of Managing Virtual Teams: Eight Key Lessons." *Harvard Management Update,* vol. 3 (1998): 4–6.

Warfield, Anne. "Do You Speak Body Language?" *Training and Development,* vol. 55 (2001): 60–62.

Weber, Max. *From Max Weber: Essays in Sociology.* Translated and edited by H. H. Gerth and C. Wright Mills. New York: Oxford University Press, 1946.

Wells, Carole V., and David Kipnis. "Trust, Dependency, and Control in the Contemporary Organization." *Journal of Business and Psychology,* vol. 15 (2001): 593–603.

Wells, Susan J. "Making Telecommuting Work." *HR Magazine,* vol. 46 (2001): 34–46.

Wheelan, Sharon A. *Creating Effective Teams.* Thousand Oaks, CA: Sage, 1999.

Wynne, Joe. "The Care and Feeding of Virtual Teams," 2000. Available: *http://myplanview.com/expert10.asp.*

Ziegler, Rene, Michael Diehl, and Gavin Zijlstra. "Idea Production in Nominal and Virtual Groups: Does Computer-Mediated Communication Improve Group Brainstorming?" *Group Processes and Intergroup Relations,* vol. 3 (2000): 141–158.

Zielinski, Dave. "Odd Language Myths: What You Think You Know about Body Language May Be Hurting Your Career." *Presentations,* vol. 15 (2001): 36, 7p.

INDEX